Plant Growth and Leaf-Applied Chemicals

Editor

Peter M. Neumann
Associate Professor and Plant Physiologist
Faculty of Agricultural Engineering
Technion — Israel Institute of Technology
Haifa, Israel

CRC Press
Taylor & Francis Group
Boca Raton London New York

CRC Press is an imprint of the
Taylor & Francis Group, an **informa** business

First published 1988 by CRC Press
Taylor & Francis Group
6000 Broken Sound Parkway NW, Suite 300
Boca Raton, FL 33487-2742

Reissued 2018 by CRC Press

Library of Congress Cataloging-in-Publication Data

Plant growth and leaf-applied chemicals.

　　Bibliography: p.
　　Includes index.
　　1 . Plant regulators--Physiological effect.
2. Plants, Effect of agricultural chemicals on.
3. Crops--Physiology.　　4. Growth (Plants)　　5. Plant
regulators--Foliar application.　　6. Agricultural
chemicals--Foliar application.　　I. Neumann, Peter M.
SB128.P5　　1988　　　　631.8　　　　87-20947
ISBN 0-8493-5414-5

A Library of Congress record exists under LC control number: 87020947

ISBN 13: 978-1-315-89663-2 (hbk)
ISBN 13: 978-1-351-07573-2 (ebk)

Visit the Taylor & Francis Web site at http://www.taylorandfrancis.com and the
CRC Press Web site at http://www.crcpress.com

PREFACE

The interactions between exogenous chemicals and the leaves of agricultural crops provide a major route by which plant development and the harvesting of food and fiber yields can be influenced. However, the chemical control of plant development is complex, and involves numerous variables, each of which can affect expected plant responses. Thus, agrochemicals and atmospheric pollutants may interact with plant leaves as droplets of aqueous solutions, oil emulsions, volatile liquids, particulate suspensions, powders, or gases; moreover, agrochemicals may be directed at the crop leaves using ground-based or air-borne equipment. Finally, plant responses are influenced by environmental conditions, the physicochemical state of the chemical at the leaf surface, and the physical, physiological, and biochemical barriers presented by the leaf cuticle and underlying tissues. Despite these difficulties, immense progress has been made in this area in recent years. The aim of this volume is to provide a compendium of state of the art overview chapters by leading researchers, from diverse scientific fields, who share a common involvement in understanding and utilizing the interactions between chemicals and plant leaves.

I wish to express my appreciation to the other chapter authors for their excellent contributions and exemplary patience while this book was being assembled.

Thanks are due to the Faculty of Agricultural Engineering for providing facilities necessary for the completion of this project. Finally, a note of appreciation to Rosy Bortman for her excellent secretarial help and to my wife and daughters for continuous and much needed support and encouragement.

THE EDITOR

Peter M. Neumann, Ph.D., is Associate Professor and plant physiologist in the Faculty of Agricultural Engineering of the Technion — Israel Institute of Technology, Haifa. He received his B.Sc. and M.Sc. degrees in biochemistry and his Ph.D. in plant physiology and biochemistry at Kings College, University of London, in 1970.

Dr. Neumann is a member of the Botanical Society of Israel, the American Society for Plant Physiology, and the American Societies for Crop Science and Agronomy. He has served as former head of the Soils and Fertilizers Research Division at the Technion and is currently an Associate Editor of the Israel Journal of Botany. He has presented many papers at international meetings and is author or co-author of more than 40 research or review articles, book chapters, and patents. His current research interests are in the hormonal and nutritional control of plant development, particularly leaf aging and senescence.

CONTRIBUTORS

Randy M. Beaudry
Research Assistant
Department of Horticulture
University of Georgia
Athens, Georgia

André Chamel
Senior Researcher
Department de Recherche Fondamentale
Centre d'Etudes Nucleaires
Grenoble, France

C. Dean Dybing
Research Plant Physiologist
U.S. Department of Agriculture
Agricultural Research Service
Plant Science Department
South Dakota State University
Brookings, South Dakota

Herbert Zvi Enoch
Senior Scientist
Department of Agricultural Meteorology
Agricultural Research Organization
Volcani Center
Bet-Dagan, Israel

John J. Hanway
Professor
Agronomy Department
Iowa State University
Ames, Iowa

Stanley J. Kays
Professor
Department of Horticulture
University of Georgia
Athens, Georgia

Charles Lay
Professor
Plant Science Department
South Dakota State University
Brookings, South Dakota

Peter M. Neumann
Associate Professor
Faculty of Agricultural Engineering
Technion — Israel Institute of
  Technology
Haifa, Israel

Daphne J. Osborne
Visiting Professor
Department of Plant Sciences
University of Oxford
Oxford, England

Lawrence O. Roth
Professor Emeritus
Department of Agricultural Engineering
Oklahoma State University
Stillwater, Oklahoma

Steven Alan Weinbaum
Professor
Pomology Department
University of California at Davis
Davis, California

TABLE OF CONTENTS

INTRODUCTION

AGROCHEMICALS: PLANT PHYSIOLOGICAL AND AGRICULTURAL
PERSPECTIVES

Peter M. Neumann

TABLE OF CONTENTS

I. AGRICULTURAL CHEMICALS AND PLANTS

The large-scale availability of industrially produced agrochemicals has helped to revolutionize agricultural practice by making it less labor intensive and by increasing yield potentials. Unfortunately, careless disposal of associated industrial wastes and improper application of agrochemicals in the field have sometimes caused adverse environmental and health problems. As in all fields of activity the potential for good and harm is inherent in the use of agrochemicals. An understanding of application strategies and the biochemical, physiological, and environmental controls regulating the interaction between plants and chemicals will ensure their safe, efficient, and economical utilization. Table 1 categorizes some of the important groups of chemicals which interact with leaves and influence plant development.

The aim of this chapter is to introduce readers from varied backgrounds to some of the roles played by the plant and its environment in either facilitating or obstructing the expected responses to foliar applications of agricultural chemicals. In order to provide a wider perspective on the foliar application of chemicals, various solute transport pathways available inside the plant and the comparative merits of other routes for introducing chemicals into plants will be considered.

II. PATHWAYS FOR SOLUTE TRANSPORT IN PLANTS

In many cases an agrochemical is required to penetrate into the cell network (symplast) in order to influence plant development. Moreover, the effectiveness of such a chemical is often proportionately greater if it is transported from the point of entry into the plant to other parts of the plant (i.e., a systemic chemical). In this section the structure and function of various long- and short-distance transport systems inside the plant will be briefly reviewed. Readers who are already well acquainted with plant transport systems could proceed directly to Section III.

A. Long-Distance Transport

1. Xylem

The xylem is the principal water-conducting vascular tissue of the plant and consists of bundles of long (up to 15 mm) lignin-reinforced, hollow cellulose tubes with tapered ends. The xylem bundles form a continuum reaching from the roots to the tips of the uppermost leaves. Specialized ray cells provide lateral transport of xylem solution in woody stems. The vascular bundles anastomise in the leaf blades so that virtually every cell is supplied by a major or minor branch vein. The typical arrangement of xylem fibers and associated cells in a vascular bundle are shown schematically in Figure 1. For detailed studies of transport tissue anatomy, readers should consult one of several excellent textbooks of plant anatomy.[22,23] The rate of delivery of xylem-transported solutes to individual leaves may change during plant aging. Ongoing changes in xylem flux of water, mineral nutrients, and plant hormones into the leaves have been implicated in the regulation of leaf development.[24,25] Flux through the xylem is regulated by root output and driven by water potential differences between soil, leaf, and atmosphere. Water flowing through the xylem is therefore usually under tension (negative pressure).[26] The distribution of solutes to different leaves is affected by ion exchange, lateral diffusion, transfer cell activity, xylem-phloem interchanges, and the hydraulic architecture of the xylem.[27] Often, the free passage of basic molecules, such as amides through the xylem, is retarded by ion-exchange binding to the xylem walls; neutral and acidic molecules may travel more freely by mass flow into the leaves.[28-30] The distribution pattern of xylem-borne solutes is of considerable importance when regulatory chemicals coming from the roots or stem are to be distributed to shoot tissues. However, since xylem

Table 1
CATEGORIES OF CHEMICALS WHICH INTERACT WITH PLANT LEAVES AND AFFECT PLANT DEVELOPMENT

A. Atmospheric nutrients and pollutants
    Sulfur dioxide, nitric oxides, acid rain, ethylene, ozone, fluorides, salt
       sprays, carbon dioxide, oxygen, and ammonia[1-5]
B. Fertilizer solutions
    Macronutrients, micronutrients, and organic nutrients[6-9]
C. Pest, disease, and weed control agents
    Biological and chemical pesticides, fungicides, bactericides, and
       herbicides[10-14]
D. Growth regulatory substances
    Growth inhibitors and stimulators, stimulators of ripening, stimulators of
       leaf and fruit abscission, dessicating agents, promoters of soil nutrient
       uptake, and photosynthesis[15-18]
E. Surface film forming compounds
    Antitranspirants, disease, and freeze protective coatings[19-21]

transport is unidirectional (into the leaves), it is of little direct importance to the subsequent transport of foliar-applied chemicals from treated leaves to rapidly growing organs with low transpiration rates, e.g., buds, flowers, or fruits. This task is carried out by the phloem.

2. Phloem

The phloem network extends from the base to the tips of higher plants and is found in association with the xylem (Figure 1). By contrast with the xylem the phloem is composed of living cells with plasma membrane and cytoplasm. Solution flow between the phloem cells occurs via the perforated sieve plates at either end of the cells and is thought to be driven by osmotically generated positive pressure gradients. The companion cells, which are an integral part of the phloem tissue, appear to provide for the metabolic requirements of the phloem cells (Figure 1).

The pH of the phloem sap is around 8.0 whereas that of xylem sap is around 6.0. Flow in the phloem network can be either acropetal (towards the top of the plant) or basipetal (towards the base). Thus, phloem transport is extremely important for the distribution of chemicals from mature leaves to growing regions in the roots and shoots. Several excellent and detailed reviews of phloem functioning are available.[31-33] Tyree et al.[34] have suggested a mathematical model which predicts phloem mobility of exogenous chemicals (xenobiotics) as a function of their membrane permeability coefficients, the dimensions of the phloem loading region, length of the translocation pathway, and phloem velocity.

B. Short-Distance Cellular Transport

The essential difference between cell-to-cell transport and organ-to-organ transport via the conductive tissues is the slower rate of the former. MacRobbie's[33] calculations indicate that the flux of sucrose in the phloem (2.5 to 20 $\times$ 10^6 pmol sec^{-1} cm^{-2}) is faster than most ion fluxes across membranes (1 to 10 pmol sec^{-1} cm^{-2}). Cell-to-cell transport can take place via the plasmodesmata, by diffusion, or via the apoplast (Figure 2). Some cells showing wall ingrowths, and believed to be especially efficient in cell-to-cell transport in and out of xylem and phloem, have been named transfer cells.[35]

III. PATHWAYS FOR ENSURING INTERACTIONS BETWEEN CHEMICALS AND PLANTS

Ideally, exogenous regulatory chemicals should reach all potential target tissues in the

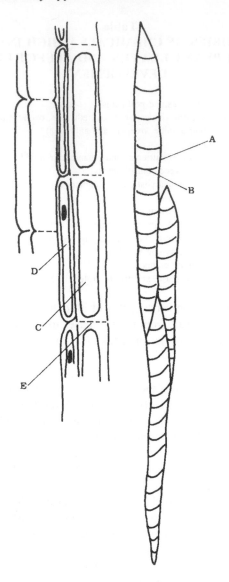

FIGURE 1.   Schematic representation of files of xylem and phloem cells. Key: A = xylem vessel; B = lignin reinforcement; C = phloem sieve tube; D = companion cell; E = sieve plate.

plant throughout the period in which developmental modification, or protection, is required. One recent and most effective approach towards obtaining this ideal situation is through the insertion of new genetic instructions to be expressed in all the cells of the engineered plant. The relative merits of this and more traditional approaches will be considered in this section.

A. Gene Insertion Pathway

An excellent example of this approach is the work of DeBlock et al.[36] who have been able to modify the plant tumor-forming plasmid of *Agrobacterium tumefaciens* so that (1) it does not induce tumors in receptor cells and (2) it can be made to carry additional exogenous, genetic capabilities (e.g., selective resistance to antimicrobial chemicals, herbicides, or pesticides) into cells. To illustrate the technique, they used a plasmid chimera

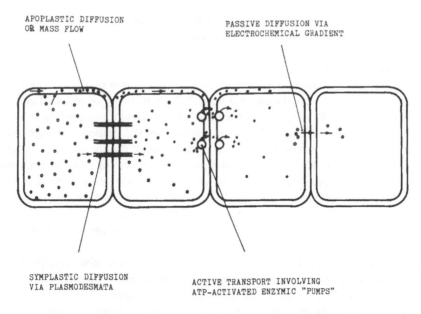

APOPLASTIC DIFFUSION
OR MASS FLOW

PASSIVE DIFFUSION VIA
ELECTROCHEMICAL GRADIENT

SYMPLASTIC DIFFUSION
VIA PLASMODESMATA

ACTIVE TRANSPORT INVOLVING
ATP-ACTIVATED ENZYMIC "PUMPS"

FIGURE 2.   Schematic representation of types of cell-to-cell transport processes.

gene to carry resistance to various antibiotics into protoplasts of tobacco cells. The phenotypically normal shoot and root tissues of whole plants regenerated from treated protoplasts had acquired antibiotic resistance. Furthermore, these plants were fully fertile and the resistant traits were subsequently transmitted in a Mendelian fashion to daughter progeny.

Clearly, this type of approach offers immense opportunities for efficient modification of plants.[37] However, many problems and limitations need to be noted before such techniques can be applied to a wide range of important agricultural crops:

1. Not all plants are as easy to culture and regenerate from protoplasts as is tobacco. However, a recent alternative method utilizing gene insertion into cells of leaf disks may prove to be more generally applicable.[38]
2. Many desirable resistance traits involve only single genes which are relatively easy to locate and manipulate. However, other traits, such as enhanced photosynthesis, delayed senescence, or increased resistance to environmental stresses and diseases, involve a multiplicity of genes, many of which will be difficult to isolate.[39] Thus, conventional selection and breeding techniques may often provide a more rapid and simple approach.[39]
3. Rates of development of crop plants are responsive to combinations of variable environmental signals (e.g., light quality and quantity, temperature, water, and nutrient status). The ability to intervene and modify plant development only after a precise state of development has been attained can be advantageous, but this may be difficult to achieve by genetic means. For example, an increased capacity for nitrogen uptake into plants might be introduced genetically.[40] Similarly, foliar sprays of some peptide alcohols are also known to promote uptake of nitrogen as ammonium and nitrate ions.[18] High levels of nitrogen uptake by plants early in the growing season are associated with lush vegetative growth so that either method of enhancing nitrogen uptake would be suitable for enhancing yields of crops required primarily for animal fodder. However, it has long been well known that excessive fertilizer-induced vegetative development in some field crops and fruit trees leads to reduced reproductive development[41] (i.e., reduced yields of grain or fruit) and/or increased harvesting difficulties. Clearly, in these cases, foliar application of a nitrogen uptake promoter at or after the initiation

of reproductive growth might be preferable to a genetically induced enhancement of nitrogen uptake throughout the growing season.

B. Soil-To-Root Pathway

The most conventional method for introducing chemicals into plants is to apply them to the soil for subsequent root uptake into the plant.[42] Plant recovery of soil-applied chemicals such as fertilizers generally ranges between 30 to 90% (J. Hagin, personal communication). The main comparative advantage of the soil pathway is the immense storage and buffer capacity of the soil. This means that large amounts of bulk chemicals, such as fertilizers, can be applied so as to provide a source of nutrients for the expanding root network throughout the growing season. Alternatively, the introduction of drip irrigation in more arid regions means that soluble chemicals can be applied to the soil-root network at selected times, compositions, and concentrations in response to plant demand.

Among the disadvantages associated with the soil pathway of application are

1.   Soil-induced chemical changes such as production of insoluble precipitates or irreversible fixation onto soil particles may make some chemicals unavailable to the plant.
2.   Leaching of applied chemicals by irrigation or rain water to depths below the reach of the plant root network can reduce uptake and also contribute to problems of groundwater pollution.
3.   The physical and chemical heterogeneity of the soil and its effectively infinite volume in the field often make it an especially unsuitable pathway for supplying low, defined dosages of growth regulatory chemicals to plants. Moreover, competitive uptake and/ or metabolism by soil microbes can significantly reduce the chemical activity available for plant uptake.
4.   Large molecules such as proteins or polysaccharides do not penetrate into the plant roots.

C. Trunk Pathway

The effectiveness of chemical formulations painted or sprayed onto the trunks or dormant branches of trees has often been investigated. Such bark dressings have been used with varying success to supply minerals,[43,44] herbicides,[45,46] pesticides,[47-49] and growth regulators[50-52] to trees.

Trunk injections have been used with varying success to introduce water soluble chemicals into trees by gravity flow or solution pressure.[49,53-55] Solid trunk implants aimed at providing a slow release of nutrients have also been tried.[56,57]

The early history and development of plant injection up to the 1930s was comprehensively reviewed by Roach.[56] He quotes pioneering reports dated 1158 and subsequent work by Leonardo da Vinci on trunk injection of arsenic for preparing poison fruit (presumably for subsequent practical use . . .). Zimmerman[58] has briefly reviewed some practical problems and solutions involved in obtaining successful distribution of liquids injected into trees.

The advantages of trunk banding or injection are

1.   Avoiding microbial, chemical, and physical inactivation associated with soil application
2.   Avoiding many pollution problems associated with soil and spray (see below) applications
3.   Direct distribution of large molecules via the xylem (injection only)

The disadvantages are

1.   Need for heavy equipment and time-consuming preparation when injection is employed
2.   Efficiency of uptake and distribution in canopy varies considerably with type of chemical, point of application, tree species or variety, and stage of growth

3.   Approach limited to woody species

D. Foliar Pathway

The main comparative advantage of the foliar spray pathway for applying exogenous solutes to field crops is that uptake commences within hours of application and can continue for several days thereafter.[59-61]

Second, the percent uptake of applied solutes can be very high, e.g., recovery of foliar-applied ^{15}N-labeled urea reached ≥70% in field trials with soybeans[60] and fruit trees.[61] Moreover, provided no chemical incompatibility exists, more than one type of compound can be applied in a single spray application[49] (e.g., pesticides with mineral nutrients). Finally, since many problems associated with soil applications are avoided, the foliar route can be used to supplement and under certain circumstances largely replace soil applications of bulk nutrients such as fertilizers.[62,63]

Possible disadvantages associated with the leaf application pathway are

1.   Spray drift may lead to deposition of chemical on nontarget sites.[49]
2.   Limited available leaf area (e.g., in seedlings) and sensitivity to burn damage caused by applied solutes[64-66] greatly limit the amount of chemical applied in one spray application and often necessitate costly and time-consuming repeat applications.
3.   The penetrability of leaf cuticles may vary considerably with leaf age, environment, and plant variety, thus hampering predictions of rate or amount of spray solute uptake.[67-72]
4.   Wetting by rain, dew, or sprinkler irrigation may leach applied substances off the leaves; moreover, photo-oxidation or surface-catalyzed decomposition of sensitive organic compounds may occur prior to cuticular penetration.[73]
5.   Rates of diffusive penetration into the leaf decrease as the molecular size of the penetrating solute is increased.[74,75]
6.   Leaf-applied chemicals may not be translocated to remote target sites, e.g., roots or new growth produced after spraying.[8,66]
7.   Possibly as a result of the number of variables involved in governing the uptake of leaf-applied chemicals, the responses to foliar applications of various groups of chemicals purported to enhance crop yields directly can be very inconsistent.[76-78]

IV. EFFECTIVENESS OF LEAF-APPLIED CHEMICALS

The potential effectiveness of leaf-applied chemicals varies with the tasks they are expected to perform. Thus, the effectiveness of barrier chemicals active at the leaf surface (e.g., film-forming or light-reflecting antitranspirants,[20,21,68] toxic protein-based insecticides,[10] or frost-protectant foam coverings[19]) is primarily a function of shoot area initially covered, durability of the covering, and rate of production of new (unprotected) shoot growth by the plant. Penetration of the leaf cuticle by such substances may cause undesirable side effects.[68]

By contrast, herbicides,[13,14] nutrients,[6-9] plant growth regulators,[15-18] systemic pesticides or microbicides,[11-14] carbon dioxide,[4] and atmospheric pollutants[1-3,5] can only affect plant growth after penetration into the cell network (symplast). The possible penetration barriers in the pathway between the waxy cuticular surface and the underlying symplasm are shown diagrammatically in Figure 3. Of these, the cuticle is believed to offer the greatest resistance.[67] Passive diffusion is probably responsible for most of the penetration of exogenous chemicals through the cuticle and underlying membranes.[64] According to Fick's first law, the rate of diffusion across a membrane is proportional to the concentration gradient across it. That is, the higher the concentration of solute which can be applied to a leaf surface without causing damage and the longer the time it remains in an active state on the leaf surface (in solution),

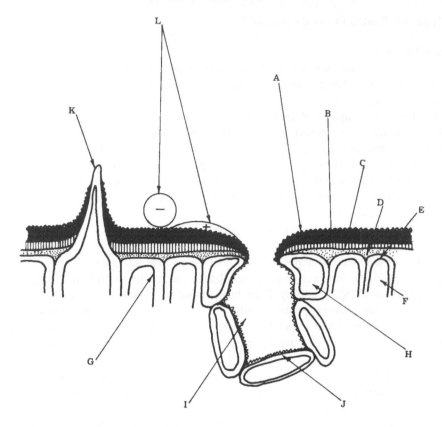

FIGURE 3.   Schematic representation of surface and interior arrangement of barriers to solute penetration.[79-81] Key: A = surface wax; B = cutin + wax; C = cutin + wax + polysaccharides; D = pectin; E = cellulose; F = epidermal cell; G = plasmalemma membrane; I = stomatal activity; J = thin wax lining; K = trichome (surface hair); L = spray droplet with (+) and without (−) surfactant.

the greater the likely rate and amount of penetration. The detailed structure and functions of plant surfaces and their penetration by various classes of solutes have been extensively and thoroughly reviewed in recent years.[9,12,74,79-81] The salient point to bear in mind when considering foliar penetration of solutes is that plant cuticles are apparently traversed by polar pathways which are permeable to water and small solute molecules; the nature and length of these polar diffusive pathways are unclear.[79,81-83] Nobel[84] has calculated that the time taken for a small molecule in a 10-mM solution to diffuse a short distance (50 μm — equivalent to the thickness of a cell) in an aqueous media would be about 0.6 sec. However, rates of diffusion of solute molecules through a plant cuticle will depend on the degree of surface wetting (Figure 3), and length (tortuosity) of the foliar pathway, and the extent of steric or ion-binding resistances.[85] Passage of radioactive ions through isolated pear leaf cuticles can usually be detected within 3 hr. (A. Chamel, personal communication).

Following cuticular penetration, uptake of solutes into the cell interior depends on the plasma membrane permeability coefficient of the molecule, the degree of cell-mediated active uptake (or rejection) and most importantly, the electrochemical concentration gradient from the outside into the cell.[9,84]

The degree of plant response to foliar-applied chemicals will depend on (1) the rate and duration of the plant uptake process and (2) the period during which an effective concentration can be maintained at the target site. Some of the extracellular factors which can limit the rate and duration of uptake of leaf-applied chemicals, together with possible practical approaches to increasing effectiveness, are indicated in Table 2.

Table 2
FACTORS AFFECTING EFFICIENT UPTAKE OF LEAF-APPLIED
CHEMICALS IN AGRICULTURAL PRACTICE

Problem	Possible solutions
Spray drifts away from target plants	Spray when wind speeds are low
	Increase droplet size[49]
Poor coverage of leaves inside plant canopy	Optimize sprayer technology
	Use larger spray volumes and high pressures[12,49]
Poor wetting of individual leaves	Improve formulation by addition of surfactant adjuvants[12,86]
Poor retention of spray by leaf surfaces	Decrease spray droplet size[87]
	Increase solution viscosity and leaf binding by adding polymeric stickers by spray solution (often has little effect on efficiency)[12,88,89]
Rapid drying of solution on leaf surface initially increases effective concentrations and accelerates penetration; subsequent drying and crystallization of (inactivated) residues on leaf surface inhibits further penetration	Spray at time of day when lowered temperature, windspeed, and high relative humidity are prevalent
	Select hygroscopic chemicals, add oils, or humectants such as glycerol, to spray solution in order to inhibit complete drying[12,90,91]
Poor cuticular penetration	Add low surface tension surfactants to encourage surface wetting and penetration of aqueous spray solutions via stomatal openings[59,86,91,92] *(N.B.:* increased run off, more rapid drying of thin surface film, and increased danger of rapid penetration leading to burn damage, may limit the desirable effects of surfactants such as L77 which encourage spontaneous spreading over leaf surfaces and stomatal penetration[91]
Single spray application at effective concentration causes unacceptable necrotic burn damage	Pretest to determine threshold concentration for onset of burn damage
	Apply lower concentrations at frequent intervals if necessary
	Higher concentrations might be applicable in encapsulated, wettable powder or other controlled release formulations, which may slow the release of active penetrant compounds on the leaf surface[49,93,94]

The cellular factors which may limit the effectiveness of leaf-applied chemicals and which are generally less subject to external control are

1. Limited duration of effect due to rapid cellular inactivation of applied chemical by degradation, conjugation, precipitation, or isolation in membrane-bound storage organelles
2. Limited transfer of applied chemical away from cells underlying sites of deposition more distant to target tissues (less of a problem when systemic [xylem and phloem mobile] chemicals are used)
3. Limitations in number or response of target or trigger sites on or inside the cells (often a function of plant developmental stage)

V. CONCLUSIONS

Clearly, the term "spray and pray" used in reference to foliar treatments of plants with chemicals is not always unfounded. A large number of quantitatively undefined and often uncontrollable variables are involved in determining the deposition, leaf penetration, and plant activity of an exogenous chemical. Nevertheless, significant progress has been made in understanding the nature of the environmental, technological, chemical, physical, and physiological factors which can be utilized to lessen the degree of dependence on prayer for ensuring adequate responses to leaf-applied chemicals. Some examples of the progress being made in understanding and exploiting the potential of the foliar route of chemical application are given in the following chapters.

REFERENCES

1. **Heath, R. L.,** Initial events in injury to plants by air pollutants, *Annu. Rev. Plant Physiol.,* 31, 395, 1980.
2. **Rennenberg, H.,** The fate of excess sulphur in higher plants, *Annu. Rev. Plant Physiol.,* 35, 121, 1984.
3. **Koziol, M. S. and Whatley, F. R., Eds.,** Gaseous air pollutants and plant metabolism, in *Proc. Int. Conf. Oxford, U.K. 1982,* Butterworths, London, 1984.
4. **Wittwer, S. H.,** Carbon dioxide levels in the biosphere: effects on plant productivity, *Crit. Rev. Plant Sci.,* 2, 171, 1985.
5. **Dabney, S. M. and Bouldin, D. R.,** Fluxes of ammonia over an alfalfa field, *Agron. J.,* 77, 572, 1985.
6. **Wittwer, S. H. and Bukovac, M. J.,** The uptake of nutrients through leaf surfaces, in *Handbuch der Pflanzenenahrung und Dungung,* Linser, H., Ed., Springer-Verlag, Berlin, 1964, 235.
7. **Mortvedt, J. J., Giordano, P. M., and Lindsay, W. L., Eds.,** *Micronutrients in Agriculture,* Soil Science Society of America, Madison, Wis., 1972.
8. **Kannan, S.,** Mechanism of foliar uptake of plant nutrients: accomplishments and prospects, *J. Plant Nutr.,* 2, 717, 1980.
9. **Swietlick, D. and Faust, M.,** Foliar nutrition of fruit crops, *Hortic. Rev.,* 6, 287, 1984.
10. **Miller, L. K., Lingg, A. J., and Bulla, L. A.,** Bacterial, viral and fungal insecticides, *Science,* 219, 715, 1983.
11. **Siegel, M. R. and Sisler, H. D., Eds.,** *Antifungal Compounds,* Marcel Dekker, New York, 1977.
12. **Hull, H. M.,** Leaf structure as related to absorption of pesticides and other compounds, *Residue Rev.,* 31, 1, 1970.
13. **Evans, J. O.,** Herbicides, in *Plant Physiology,* 2nd ed., Salisbury, F. B. and Ross, C., Eds., Wadsworth Publishing, Belmont, 1978, 393.
14. **Audus, L. J.,** Herbicides, in *Physiology Biochemistry Ecology,* Vols. 1 and 2, 2nd ed., Academic Press, New York, 1977.
15. **Nickell, L. G.,** *Plant Growth Regulating Chemicals,* Vols. 1 and 2, CRC Press, Boca Raton, Fla., 1983.
16. **Ory, R. L and Rittig, F. R.,** Bioregulators — chemistry and uses, *ACS Symp. Series 257,* American Chemical Society, Washington, D.C., 1984.
17. **Manning, D. T., Campbell, A. J., Chen, T. M., Tolbert, N. E., and Smith, E. W.,** Detection of chemicals inhibiting photorespiratory senescence in a large-scale survival chamber, *Plant Physiol.,* 76, 1060, 1984.
18. **Lin, L. W. and Kauer, J. C.,** Peptide alcohols as promoters of nitrate and ammonium ion uptake in plants, *Plant Physiol.,* 77, 403, 1985.
19. **Siminovitch, D., Ball, W. L., Desjardins, R., and Gamble, D. S.,** Use of protein-based foams to protect plants against frost, *Can. J. Plant Sci.,* 47, 11, 1967.
20. **Fuchs, M., Stanhill, G., and Moreshet, S.,** Effect of increasing foliage and soil reflectivity of the solar radiation balance on wide row grain sorghum, *Agron. J.,* 68, 865, 1976.
21. **Ziv, O. and Frederickson, I.,** Control of foliar diseases with epidermal coating materials, *Plant Dis.,* 67, 212, 1983.
22. **Esau, K.,** *Anatomy of Seed Plants,* 2nd ed., John Wiley & Sons, New York, 1977.
23. **Fahn, A.,** *Plant Anatomy,* 3rd ed., Pergamon Press, Oxford, 1982.
24. **Neumann, P. M. and Stein, Z.,** Xylem transport and the regulation of leaf metabolism, *What's New in Plant Physiol.,* 14, 33, 1983.
25. **Neumann, P. M. and Stein, Z.,** Relative rates of delivery of xylem solute to shoot tissues: possible relationship to sequential leaf senescence, *Physiol. Plant.,* 62, 390, 1984.
26. **Scholander, P. F., Hammel, H. T., Bradstreet, E. D., and Hemmingsen, E. A.,** Sap pressures in vascular plants, *Science,* 148, 339, 1965.
27. **Neumann, P. M. and Noodén, L. D.,** Pathway and regulation of phosphate translocation to the pods of soybean explants, *Physiol. Plant.,* 60, 166, 1984.
28. **Charles, A.,** Uptake of dyes into cut leaves, *Nature,* 171, 435, 1953.

29. **Dickson, R. E., Vogelman, T. C., and Larson, P. R.,** Glutamine transfer from xylem to phloem and translocation to developing leaves of populus deltoides, *Plant Physiol.,* 77, 412, 1985.

30. **McNiel, D. L., Atkins, C. A., and Pate, J. S.,** Uptake and utilization of xylem-borne amino compounds by shoot organs of a legume, *Plant Physiol.,* 63, 1076, 1979.

31. **Zimmerman, M. H. and Millburn, J. H., Eds.,** Phloem transport, in *Encyclopedia of Plant Physiology New Series,* Vol. 1 (Part A), Springer-Verlag, Berlin, 1975.

32. **Cronshaw, J.,** Phloem structure and function, *Annu. Rev. Plant Physiol.,* 32, 465, 1981.

33. **MacRobbie, E. A. C.,** Phloem translocation, facts and mechanisms: a comparative survey, *Biol. Rev.,* 46, 429, 1971.

34. **Tyree, M. T., Peterson, C. A., and Edgington, L. V.,** A simple theory regarding ambimobility of xenobiotics with special reference to the nematicide oxamyl, *Plant Physiol.,* 63, 367, 1979.

35. **Gunning, B. E. S. and Pate, J. S.,** "Transfer Cells", Plant cells with wall ingrowth specialized in relation to short-distance transport of solutes — their occurrence, structure and development, *Protoplasma,* 68, 107, 1969.

36. **Deblock, M., Herrera Estrella, L., Van Montau, M., Schell, J., and Zambryski, P.,** Expression of foreign genes in regenerated plants and in their progeny, *EMBO J.,* 3, 1681, 1984.

37. **Comai, L. and Stalker, D. M.,** Impact of genetic engineering on crop protection, *Crop Prot.,* 3, 399, 1984.

38. **Horsch, R. B., Fry, J. E., Hoffmann, N. L., Eicholz, D., Rogers, S. G., and Fraley, R. T.,** A simple and general method for transferring genes into plants, *Science,* 227, 1229, 1985.

39. **Blum, A.,** Breeding crop varieties for stress environments, *Crit. Rev. Plant Sci.,* 2, 199, 1985.

40. **Hardy, W. R. F. and Havelka, U. D.,** N_2 fixation research — a key to world food, *Science,* 188, 635, 1975.

41. **Kraus, E. J. and Kraybill, H. R.,** Vegetation and reproduction with special reference to the tomato, *Oreg. Agric. Exp. Stn. Bull.,* 149, 5, 1918.

42. **Russel, E. W.,** *Soil Conditions and Plant Growth,* 10th ed., Longman, London, 1973.

43. **Tuckey, H. B., Ticknor, R. L., Hinsvark, O. N., and Wittwer, S. H.,** Absorption of nutrients by stems and branches woody plants, *Science,* 116, 167, 1952.

44. **Neumann, P. M.,** Supply of micronutrients to apple trees via the bark, *Commun. Soil Sci. Plant Anal.,* 10, 1481, 1979.

45. **Sundaram, A.,** A preliminary investigation of the penetration and translocation of 2,4,5-T in some tropical trees, *Weed Res.,* 5, 213, 1965.

46. **Upchurch, R. P., Keaton, J. A., and Cobble, H. O.,** Response of turkey oak to 2,4,5-T as a function of final formulation oil content, *Weed Sci.,* 17, 505, 1969.

47. **Jeppson, L. R., Jesser, M. J., and Complin, J. O.,** Tree trunk application as a possible method of using systemic insecticides on citrus, *J. Econ. Entomol.,* 45, 669, 1952.

48. **Coppel, H. C. and Norris, D. M.,** Bark penetration and uptake of systemic insecticides from several treatment formulations in white pines, *J. Econ. Entomol.,* 59, 928, 1966.

49. **Mathews, G. A.,** *Pesticide Application Methods,* Longman, New York, 1982.

50. **Backhaus, R. A. and Sachs, R. M.,** Tree growth inhibition by bark application of morphactins, *Hortic. Sci.,* 11, 578, 1976.

51. **Neumann, P. M., Backhaus, R., Doss, R. P., and Sachs, R. M.,** Site of in-vivo regulation of tree stem elongation by bark-banded morphactins, *Physiol. Plant.,* 49, 55, 1977.

52. **Neumann, P. M., Doss, R. P., and Sachs, R. M.,** A new laboratory method used for investigating the uptake translocation and metabolism of bark banded morphactins by trees, *Physiol. Plant.,* 39, 248, 1977.

53. **Ketchie, D. O. and Williams, M. W.,** A method for feeding chemicals into young apple trees, *Hortic. Sci.,* 7, 491, 1972.

54. **Worley, R. E., Littrell, R. H., and Polles, S. G.,** Pressure trunk injection promising for pecan and other trees, *Hortic. Sci.,* 11, 590, 1976.

55. **Jaynes, R. A. and Van Alfen, N. K.,** Control of American chestnut blight by trunk injection with methyl-2-benzimidazole carbonate M.B.C., *Phytopathology,* 64, 1479, 1974.

56. **Roach, W. A.,** Plant injection as a physiological method, *Ann. Bot. (London),* 3, 155, 1939.

57. **Nevrot, J. and Banin, A.,** Trunk in planted zinc bentonite as a source of zinc for apple trees, *Plant Soil,* 69, 85, 1982.

58. **Zimmerman, M. H.,** *Xylem Structure and the Ascent of Sap,* Springer-Verlag, Berlin, 1983, 123.

59. **Weinbaum, S. A. and Neumann, P. M.,** Uptake and metabolism of ^{15}N-labeled potassium nitrate by French Prune *(Prunus domestica* L.) leaves and the effects of two surfactants, *J. Am. Soc. Hortic. Sci.,* 102, 601, 1977.

60. **Vasilas, B. L., Legg, J. O., and Wolf, D. C.,** Foliar fertilization of soybeans: absorption and translocation of ^{15}N-labelled urea, *Agron. J.,* 72, 271, 1980.

61. **Klein, I. and Weinbaum, S. A.,** Foliar retention of urea to almond and olive: retention and kinetics of uptake, *J. Plant Nutr.,* 8, 117, 1985.

62. **Embleton, T. W. and Jones, M. W.**, Foliar-applied nitrogen for citrus fertilization, *J. Environ. Qual.*, 3, 388, 1974.

63. **Weinbaum, S. A.**, Feasibility of satisfying total nitrogen requirement of non-bearing prune trees with foliar nitrate, *Hortic. Sci.*, 13, 52, 1978.

64. **Neumann, P. M., Ehrenreich, Y., and Golab, Z.**, Foliar fertilizer damage to corn leaves relation to cuticular penetration, *Agron. J.*, 73, 979, 1981.

65. **Neumann, P. M. and Golab, Z.**, Comparative effects of mono- and dipotassium phosphates on cell leakiness in corn leaves, *J. Plant Nutr.*, 6, 275, 1983.

66. **Neumann, P. M. and Prinz, R.**, Foliar iron spray potentiates growth of seedlings on iron-free media, *Plant Physiol.*, 55, 988, 1975.

67. **Leece, D. R.**, Composition and ultrastructure of leaf cuticles from fruit trees relative to differential foliar absorption, *Aust. J. Plant Physiol.*, 3, 833, 1976.

68. **Naim, R. and Neumann, P. M.**, Mechanism of senescence induction by silicone oil, *Physiol. Plant.*, 42, 57, 1978.

69. **Bengston, C., Larsson, S., and Liljenberg, C.**, Effects of water stress on cuticular transpiration rate and amount and composition of epicuticular wax in seedlings of six oat varieties, *Physiol. Plant.*, 44, 319, 1978.

70. **Hull, H. M., Went, F. W., and Bleckmann, C. A.**, Environmental modification of epicuticular wax structure *Prosopis* leaves, *J. Ariz.- Nev. Acad. Sci.*, 14, 39, 1979.

71. **Darnell, R. L. and Ferree, D. C.**, The influence of environment on apple tree growth wax formation and foliar absorption, *J. Am. Soc. Hortic. Sci.*, 108, 506, 1983.

72. **Bukovac, M. J., Flore, J. A., and Baker, E. A.**, Peach leaf surfaces: changes in wetability retention, cuticular permeability and epicuticular wax chemistry during expansion with special reference to spray application, *J. Am. Soc. Hortic. Sci.*, 104, 611, 1979.

73. **Sachs, R. M., Ryugo, K., and Messerschmidt, O.**, Non-metabolic catalysis of ^{14}C furfuryl amino purine (kinetin) on plant surface, glass and porcelain ware, *Plant Physiol.*, 57, 98, 1976.

74. **Cutler, D. F., Alvin, K. L., and Price, C. E.**, *The Plant Cuticle*, Linnean Soc. Symp. Ser., Academic Press, London, 1982.

75. **McFarlane, J. C. and Berry, W. L.**, Cation penetration through isolated leaf cuticles, *Plant Physiol.*, 53, 723, 1973.

76. **Abetz, P.**, Seaweed extracts: have they a place in Australian agriculture or horticulture?, *J. Aust. Inst. Agric. Sci.*, 46, 23, 1980.

77. **Neumann, P. M.**, Late-season foliar fertilization with macronutrients — is there a theoretical basis for increased seed yields?, *J. Plant Nutr.*, 5, 1209, 1982.

78. **Ries, S. K.**, Regulation of plant growth with triacontanol, *Crit. Rev. Plant Sci.*, 2, 239, 1985.

79. **Franke, W.**, Mechanisms of foliar penetration of solutions, *Annu. Rev. Plant Physiol.*, 18, 281, 1967.

80. **Kolattukudy, P. E.**, Biosynthesis of cuticular lipids, *Annu. Rev. Plant Physiol.*, 21, 163, 1970.

81. **Juniper, B. E. and Jeffree, C. E.**, *Plant Surfaces*, Edward Arnold, London, 1983.

82. **Schönherr, J. and Bukovac, M. J.**, Preferential polar pathways in the cuticle and their relationship to ectodesmata, *Planta*, 92, 189, 1970.

83. **Haas, K. and Schönherr, J.**, Composition of soluble cuticular lipids and water permeability of cuticular membranes from citrus leaves, *Planta*, 146, 399, 1979.

84. **Nobel, P. S.**, *Plant Cell Physiology: A Physicochemical Approach*, W. H. Freeman, San Francisco, 1970, 7.

85. **Schönherr, J. and Huber, R.**, Plant cuticles are polyelectrolytes with isoelectric points around three, *Plant Physiol.*, 59, 149, 1977.

86. **Neumann, P. M. and Prinz, R.**, Evaluation of surfactants for use in the spray treatment of iron chlorosis in citrus trees, *J. Sci. Food Agric.*, 25, 221, 1974.

87. **Blackman, G. E., Bruce, R. S., and Holly, K.**, Studies in the principles of phytotoxicity. V. Interrelationships between specific differences in spray retention and selective toxicity, *J. Exp. Bot.*, 9, 175, 1958.

88. **Sommers, E.**, Formulation, in *Funghicides: An Advanced Treatise*, Vol. 1, Torgeson, Ed., Academic Press, New York, 1967, 153.

89. **Ogawa, J. M., MacSwan, I. C., Manji, B. T., and Schick, F. J.**, Evaluation of funghicides with and without adhesives for control of peach diseases under low and high rainfall, *Plant Dis. Rep.*, 61, 672, 1977.

90. **Reed, D. W. and Tukey, H. B., Jr.**, Effect of pH on foliar absorption of phosphorus compounds by *Chrysanthemum*, *J. Am. Soc. Hortic. Sci.*, 103, 337, 1978.

91. **Leece, D. R. and Dirou, J. F.**, Comparison of urea foliar sprays containing hydrocarbon or silicon surfactants with soil-applied nitrogen in maintaining the leaf nitrogen content of prune trees, *J. Am. Soc. Hortic. Sci.*, 104, 644, 1979.

92. **Schönherr, J. and Bukovac, M. J.**, Penetration of stomata by liquids: dependence on surface tension wettability and stomatal morphology, *Plant Physiol.*, 49, 813, 1972.

93. **Barel, D. and Black, C. A.,** Foliar application of P screening of various inorganic and organic P compounds, *Agron. J.,* 71, 15, 1979.
94. **Barel, D. and Black, C. A.,** Foliar application of P II yield responses of corn and soybean sprayed with various condensed phosphates and P-N compounds in greenhouse and field experiments, *Agron. J.,* 71, 21, 1979.

Chapter 1

APPLICATION TECHNOLOGY

Lawrence O. Roth

TABLE OF CONTENTS

I. INTRODUCTION

A rather extensive technology surrounds the application of chemicals to plants. Unfortunately, this technology is not fully developed as it has not had the resources and scientific effort behind it as have the disciplines and industries developing and producing the chemicals. Only relatively recently have scientific efforts been intensified to develop the necessary technologies to enable precision in application — to be able to apply the chemicals in the quantity and manner that they will be most effective.

Today, the general notion that existing application equipment is expected to efficiently, effectively, and safely apply all past, present, and future chemical compounds and formulations presented to them is certainly obsolete. The rather sophisticated (and costly) chemicals being developed and used certainly demand new concepts to apply them precisely in a prescribed manner. It is quite likely, then, that developments in application equipment and techniques may well be more product specific and/or application technique specific to take full advantage of the properties and characteristics of the chemicals and the surfaces to which they will be applied.

This introductory chapter will deal largely with a brief description of the state-of-the-art of application technology and, hopefully, will provide a better understanding of the potentials and limitations involved. For those desiring to delve deeper into this technology, several excellent references[1-3] will provide valuable information and leads to other sources.

II. ATOMIZATION

A. The Process

Atomization is the process of creating a system of drops from a mass of liquid. Briefly, the process involves the formation and disintegration of liquid filaments and thin liquid sheets as surface tension is overcome by turbulence and other forces when the liquid emerges from a confined state. Actual drop formation occurs over a very short interval of time, a microsecond or less, and generally within a few centimeters of the atomizer. The nature of the atomization process and the resulting drop size distributions will be largely a function of the type of atomizer and its operating conditions as well as the principal liquid physical properties of viscosity, surface tension, and density. The process results in a great increase in the surface area of the liquid exposed to the environment. With the commercial atomization equipment now available, a range of drop sizes will be generated. Several types of devices are used to atomize and apply chemicals to plants.

B. Pressure Atomizers

The most commonly available and widely used device is the pressure atomizer where liquid under pressure is forced through an orifice. Depending on the orifice shape, size, and nozzle internal configuration, the emerging liquid will form drop streams from one or more circular orifices, an oval or fan shaped pattern where the liquid is forced through an elliptical-shaped orifice, or a conical spray pattern from a single circular exit orifice with an internal nozzle design that imparts a tangential or swirling motion to the liquid just prior to passing through the exit orifice. In each of these cases, the spray drops are created as the result of the formation and disintegration of the emerging liquid filaments and sheets. The atomization resulting from circular orifices (Figure 1) may produce relatively large uniform drops (very nearly twice the diameter of the orifice) interspersed with occasional small drops. However, unless some means of dispersing the stream over a greater width, such as a multiple orifice nozzle, a mechanical device such as the Vibrajet,[4] or an electrostatic scheme described by Roth and Porterfield[5] are employed to accomplish stream dispersion, the utility of single drop streams is quite limited. The other pressure nozzle orifice configurations also produce

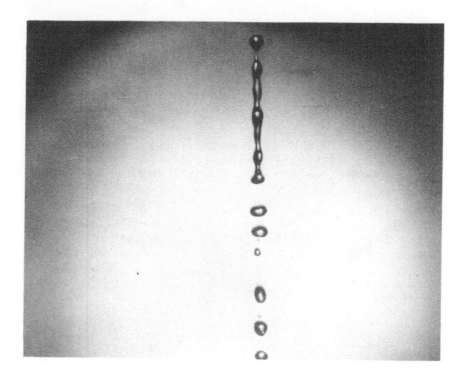

FIGURE 1. The atomization of a jet stream of water emerging from a circular orifice at 7 kPa.

directional sprays that are dispersed over greater areas (Figure 2). All of these sprays will feature a range of drop sizes, the exact distribution depending on the orifice (and internal) design, operating pressure, and liquid physical properties (Figure 3). The sprays from pressure atomizers are not only directional, but the drops will possess a velocity (related to pressure) that will affect the distance traveled as well as the impact forces upon contact with a surface.

C. Pneumatic Atomizers

Pneumatic atomization occurs where a pressurized air stream impinges on a liquid jet stream in or near the nozzle exit (Figure 4) and causes it to disintegrate into a system of drops. Pneumatic atomizers are characterized by their ability to produce sprays of very small drops and hence have found application mostly in ultra low volume (ULV) applications (2 ℓ/ha or less) where very low liquid flow rates and small drop sizes are required. Systems involving pneumatic atomization tend to be more complicated as a regulated air system is required in addition to the liquid system.

D. Spinning Disk Atomization

Spinning disk atomization consists of centrifugally accelerating a liquid on the surface of a rotating disk and allowing it to escape tangentially in a sheet or a series of filaments around the disk periphery that subsequently disintegrate into a system of drops (Figure 5). These atomizers will produce a range of drop sizes, though the drop spectrum will be much narrower than that produced by pressure atomizers.[6] The principal equipment variables influencing drop formation are the liquid flow rate and the disk rotative speed. In addition, the basic liquid physical properties of surface tension, viscosity, and density influence the drop formation process. The acronym CDA (controlled drop application) is commonly associated with a particular device employing this method of atomization and is based on the premise that some control over the atomization process is occurring.

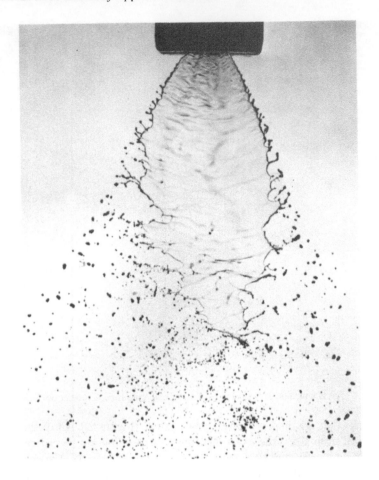

FIGURE 2.   The atomization of water from a pressure nozzle with an oval-shaped orifice at 140 kPa.

E. Electrodynamic and Electrostatic Atomization

An electrodynamic atomizing nozzle, the Electrodyn®, uses high voltage to atomize liquids having unique conductivity characteristics. This electrodynamic system produces very small charged drops that result in improved coverage with a reduced drift potential (Figure 6). At present, this system is limited to a single insecticide formulation, but further development should result in other materials being available. Electrostatic charging systems used in conjunction with pressure or pneumatic atomizers, that induce charges on the surface of the drops formed to achieve dispersion and enhance deposition, are under development, but are not commercially available at this time.[8]

F. Wiping

A relatively new development for selective application without atomization involves a wiping or rubbing technique to effect the transfer of chemicals.[9] Here absorbent materials are continuously wetted and saturated with the chemical solution that are then wiped onto target surfaces when the saturated material makes contact with the target. The major limitations of these systems are appropriate configurations of the absorbent surfaces to insure adequate and sufficient contact with the targets.

G. Roller-Brush Atomization

A system of spray drops are formed from a roller-brush unit (Figure 7) when the bristles

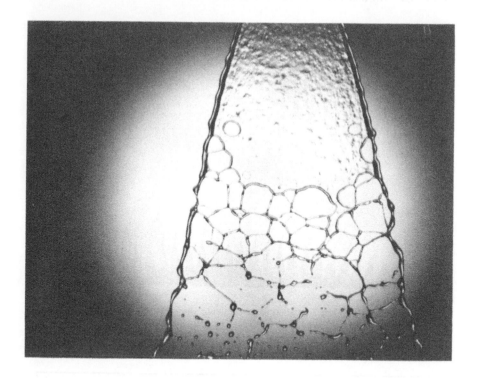

FIGURE 3.   The atomization of a water-based viscous liquid from a pressure nozzle with an oval-shaped orifice at 70 kPa.

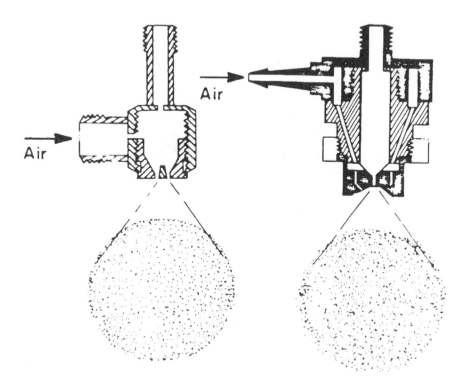

FIGURE 4.   Cross-sections of two pneumatic atomizing nozzles; an internal mixing type (left) and an external mixing type (right).

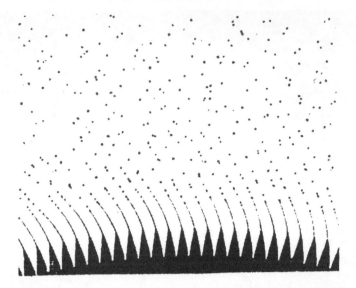

FIGURE 5. Drop formation from a spinning disk atomizer where individual filaments formed at each disk point break up into a series of relatively uniform drops.

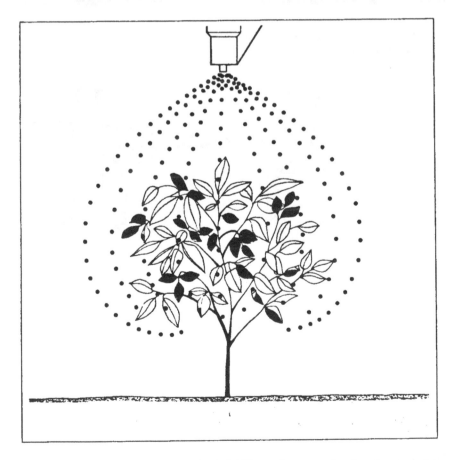

FIGURE 6. Schematic diagram of the electric field lines and corresponding drop streams created by an electrodynamic atomizing nozzle.

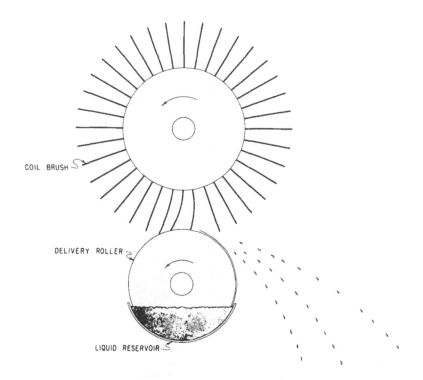

COIL BRUSH

DELIVERY ROLLER

LIQUID RESERVOIR

FIGURE 7. Schematic diagram of a roller-brush system that atomizes liquid from the surface of a wetted delivery roller.

of a rotating brush contact the wetted surface of a roller and remove liquid from the surface of the roller. Drops are formed by the whipping action of the bristles as the bristles lose contact with the roller and move through the thin layer of liquid on the roller surface. Test work with such a system[10] produced a low-velocity directional spray composed of a wide range of drop sizes.

H. Uniform Drop Generators

The goal of most atomization researchers has been to develop practical means to generate sprays composed of uniform drops of a predictable size. Unfortunately, only a few successes are to be noted and, at the present time, commercial equipment to accomplish this task is not available. Uniform drop generating systems involving piezoelectric and magnetostrictive devices (Figure 8) were developed,[11] patented,[12] and used in field experiments.[13,14] Other researchers[15,16] have more recently made and tested adaptations of these devices, but also without commercial success. Thus, the goal remains elusive and the equipment now available is not much beyond the category of interesting and perhaps useful laboratory devices.

I. Spray Distribution

Where plant research involves the use of atomizers to apply chemicals, it is important to characterize the spray in some detail in order to properly evaluate this experiment parameter. One must question published work involving drop-size studies where the spray distribution parameters have been inadequately characterized or described. Accurate sampling of the drop size spectra of sprays requires an elaborate, time-consuming, and often expensive procedure. Great strides have been made in recent years in rapid and accurate means of establishing the drop size distributions of sprays using laser technology.[17,18] Sprays commonly used in biological experiments contain a range of drop sizes and can ordinarily be represented by a normal or log-normal distribution on a number or a volume basis. When such data are

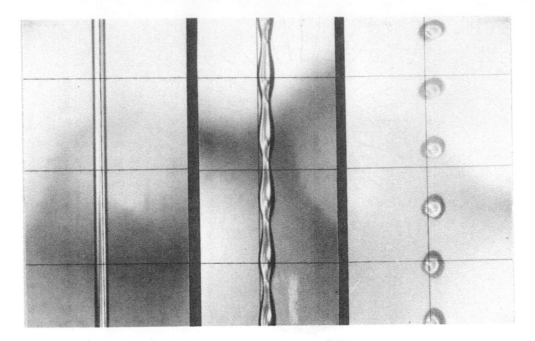

FIGURE 8.   The development of uniform drops from an undisturbed jet stream of water (left) at 276 kPa as a cyclic vibration of 18,300 Hz (center) is impressed on the liquid with a magnetostrictive device placed just upstream of the orifice. Each of the nodes in the stream will form a drop (right) that is very nearly equal in size to the others.

available, the spray can be quite completely described with a median diameter and another parameter such as the standard deviation.[19] The ASTM[20] recommends a standardized notation for the statistical parameters in spray distributions and the terms "range" and "relative span" have been defined[18] to assist in describing the uniformity of drop spectra. Quite different distributions result from different atomizers for different liquids and operating conditions. Most of the distribution data found in the literature to date have been collected while spraying water into still air.[21] More recent studies[17,18] have involved sampling the spray in a high speed air stream (simulating aircraft application) and the use of actual formulated chemicals or liquids simulating the physical properties of formulated (and diluted) chemicals. A variety of spray additives are also available that alter the surface tension, viscosity, and density and affect the atomization and drop spectra. Care must thus be exercised in selecting distribution data if the spray distribution cannot be established for the particular atomizer, liquid, and set of operating conditions.

J. Evaporation

Formulated products to be applied are frequently a mixture of several quite volatile liquids. Sprays are commonly applied at temperatures when biological activity is great and under these conditions the opportunity for significant evaporation may exist. As previously indicated, the atomizers commonly used produce a wide range of drop sizes, some with 5 to 15% of the spray (by volume) composed of drops less than 100 μm in diameter. The lifetime of the smaller drops in this part of a drop spectrum, when introduced into a typical environment, may be very short, perhaps on the order of a few milliseconds. Since the drops must travel some distance from the atomizer to the target, the smallest drops may lose a significant portion of their mass on the way to the target and indeed, if composed of relatively volatile liquids, may lose mass to a point where they are essentially unaffected by gravity and would be carried to (or away from) a target by other forces. The evaporation of moving drops is a very complex simultaneous heat and mass-transfer process and thus is difficult to quantify for a cloud of moving nonhomogeneous spray drops. The practical consideration, then, to minimize evaporation is to use (where possible) liquids of relatively low volatility, to minimize the small drop component of a spray, and to avoid (if possible) highly evaporative environmental application conditions. The use of vegetable oil diluents[22,23] is one way to reduce the effects of evaporation.

III. APPLICATION SYSTEMS

A variety of systems exist for making applications. It is important to note that, with the exceptions of the wiping and electrostatic methods of application, once atomization has occurred, control over the fate of the drops is essentially lost and the drops will travel to a target with the energy, velocity, and direction imparted to them during the atomization process. Their trajectories will be affected by the force of gravity and if the drops are small enough or evaporate enough, may also be affected by other forces in the surrounding environment. Thus, just how a given drop may approach and impact on a target surface cannot be accurately predicted or described.

A. Basic Requirements

There are several basic requirements to consider when selecting application equipment. First, several factors are work or job related. The required productivity or capacity (hectares per hour) is based on the size of the area to be treated, how often the area is to be treated, and how fast the job should be done. This consideration will establish the required size of the unit (swath width and tank size) and the number of units that may be required. The spray volume (liters per hectare) required or desired for a particular application affects the nozzle size, the pump capacity, and the amount of diluent needed (tank size). The type of terrain, slope, and ground obstacles determines whether manual or powered equipment can be used. The planting pattern and the type and size of crop and plant spacings affect sprayer operating pressure, nozzle type and size, pump size, and whether the equipment can be manual or powered. The type of coverage required for a specific application will influence the nozzle type and size and operating pressure. The type of formulation, whether an emulsifiable concentrate or a wettable powder, will affect the type and amount of tank agitation required to create and maintain a proper solution or mixture. The relative toxicity of the material will determine the methods of handling and mixing needed and the safety precautions to be taken to protect man, animals, and the environment. Other factors are machine related. Since practically all formulated products are chemically active, the materials used in the sprayers should feature corrosion resistance to provide assurance that the equipment will operate when needed. The equipment needs to be durable enough to stand up under all field operating conditions and feature reliability to operate when needed. The equipment should feature simplicity in construction and operation. It should be simple and easy to adjust, operate, and clean following use. Equipment should be purchased from companies and dealers having good reputations for providing spare parts and service.

B. Types of Equipment

The many types and sizes of equipment available range from the simplest hand sprayer to expensive and complex aircraft equipment. Each piece of equipment has been generally designed for a specific job or duty, but most can be adapted and used for more than one type of application. From the assortment of equipment available, it is possible to select appropriate equipment to accomplish most specific application tasks. The several classes of equipment follow. In each class there are several sizes to choose from, i.e., tank size, pumping capacity, nozzle size, and boom width. Many of these types of equipment are described in more detail in a recent work by Matthews.[2]

1.   Manually transported (hand operated, compressed gas operated, or motorized) equipment

    a.   Hand sprayers, continuous and intermittent
    b.   Compression, different size tanks
    c.   Back pack, manually operated or motorized
    d.   Slide pump
    e.   Barrel and wheelbarrow
    f.   Electrodynamic and electrostatic
    g.   Spinning disk
    h.   Wiping
    i.   Motorized pump with handgun

2.   Ground contact equipment (pull type, tractor mounted, or self-propelled)

    a.   Boom type for row-crop and broadcast application (pressure nozzles or spinning disks)
    b.   Boomless type for broadcast application
    c.   Handgun for tree spraying or spot treatment
    d.   Air carrier (high volume to drench foliage to run-off or low volume to apply small quantities in fine mist)
    e.   Wipe on rope-wick types

3.   Aircraft (fixed wing and helicopter)

    a.   For conventional volumes of 20 to 50 ℓ/ha or for ultra low volumes of 0.5 to 1.0 ℓ/ha

C. Proper Use

Once an application need has been identified and equipment selected, then the details of application must be satisfied. The chemicals must be properly and accurately measured, mixed, and applied in the proper manner. This involves instructions and precautions in safe handling, mixing, pouring, calibration, and application, and includes techniques of application so that the equipment is used properly. Proper use also includes care during and following use. Probably more equipment failure is due to improper or inadequate maintenance following use than for any other reason. Although many components are made out of inert materials, i.e., chemically inactive materials, this is not always possible for reasons of economy or durability. Thus the great need exists to give proper care to the system immediately following use to minimize the amount of corrosion and other damage that may result due to prolonged exposure of the equipment parts to the chemically active materials commonly found in formulated products.

REFERENCES

1. **Marshall, W. R., Jr.,** Atomization and spray drying, *Chem. Eng. Prog. Monogr. Ser. No. 2,* 50, 1954.
2. **Matthews, G. A.,** *Pesticide Application Methods,* Longman, New York, 1979.
3. **Metcalf, R. L.,** *Advances in Pest Control Research,* Vol. 2, Interscience, New York, 1958.
4. **Lake, J. R.,** Spray formation from vibrating jets, *Br. Crop Prot. Counc. Monogr. No. 2,* 61, 1970.
5. **Roth, L. O. and Porterfield, J. G.,** Liquid atomization for drift control, *Trans. ASAE,* 9, 553, 1966.
6. **Bode, L. E., Butler, B. J., Pearson, S. L., and Bouse, L. F.,** Characteristics of the Micromax rotary atomizer, *Trans. ASAE,* 26, 999, 1983.

7. Plant Protection Division, ICI Ltd., Development Bulletin: The Hand Held Electrodyn Sprayer, Surrey, England, 1981.

8. **Law, S. E.,** Embedded electrode electrostatic induction spray charging nozzle: theoretical and engineering design, *Trans. ASAE,* 21, 1096, 1978.

9. **Dale, J. E.,** Ropewick applicator — tool with a future, *Weeds Today,* 12, 3, 1980.

10. **Roth, L. O. and Tripp, G. W.,** A Roller-brush atomizer, *Trans. ASAE,* 16, 653, 1973.

11. **Roth, L. O. and Porterfield, J. G.,** Spray drop size control, *Trans. ASAE,* 13, 779, 1970.

12. **Vehe, D. A., Porterfield, J. G., and Roth, L. O.,** Jet Stream Vibratory Atomizing Device, U.S. Patent 3,679,132, 1972.

13. **Jiminez, E., Roth, L. O., and Young, J. H.,** Droplet size and spray volume influence on control of the bollworm, *J. Econ. Entomol.,* 69, 327, 1970.

14. **Buehring, N., Roth, L. O., and Santlemann, P. W.,** Fluometuron and MSMA phytotoxicity as influenced by drop size and concentration, Proc. 24th South. Weed Sci. Soc., Memphis, 1971.

15. **Bouse, L. F., Haile, D. G., and Kunze, O. R.,** Cyclic disturbance of jets to control spray drop size, *Trans. ASAE,* 17, 235, 1974.

16. **Wilce, S. E., Akesson, N. B., Yates, W. E., Christensen, P., Lowden, R. E., Hudson, D. C., and Weigt, G. I.,** Drop size control and aircraft spray equipment, *Agric. Aviat.,* 16(1), 7, 1974.

17. **Bouse, L. F. and Carlton, J. B.,** Factors affecting size distribution of vegetable oil spray droplets, ASAE Paper AA83-006, presented at the ASAE/NAAA Technical Session on Agricultural Aviation Research, Reno, 1983.

18. **Yates, W. E. and Akesson, N. B.,** Drop size spectra from nozzles in high-speed airstreams, ASAE Paper AA83-005, presented at the ASAE/NAAA Technical Session on Agricultural Aviation Research, Reno, 1983.

19. **Bode, L. E. and Butler, B. J.,** The three D's of droplet size: diameter, drift and deposit, ASAE Paper AA81-004, presented at the ASAE/NAAA Technical Session on Agricultural Aviation Research, Las Vegas, 1981.

20. American Society for Testing Materials, *Standard Practice for Determining Data Criteria and Processing for Liquid Drop Size Analysis,* ASTM Standards, 14.02, Philadelphia, 1983, 687.

21. **Tate, R. W. and Janssen, L. F.,** Droplet size data for agricultural spray nozzles, *Trans. ASAE,* 9, 303, 1966.

22. **King, E. E. and King, R.,** Droplet size versus biological effectiveness of oil sprays, ASAE Paper AA83-008, presented at the ASAE/NAAA Technical Session on Agricultural Aviation Research, Reno, 1983.

23. **Wodageneh, A. and Matthews, G. A.,** The addition of oil to pesticide sprays — effect on droplet size, *Trop. Pest Manage.,* 27, 121, 1981.

Chapter 2

FOLIAR UPTAKE OF CHEMICALS STUDIED WITH WHOLE PLANTS AND ISOLATED CUTICLES

André Chamel

TABLE OF CONTENTS

I. INTRODUCTION

A wide diversity of substances can be deposited on the aerial parts of plants either intentionally as with nutrients, growth regulators, herbicides, fungicides, and insecticides applications to promote, control, and protect plant development, or unintentionally as in the case of industrial pollutants. Therefore, the problem of foliar penetration and translocation of these chemicals is of major practical importance. It can be considered from diverse aspects such as the effects of foliar applications or the pathways and mechanisms of foliar absorption. The purpose of this review is to survey some approaches to studying the behavior of chemicals applied to leaves, as illustrated by results obtained by the author with whole plants and in vitro using isolated cuticles. The experimentation on whole plants gives an overall view of penetration and translocation and enables the influence of various factors on each of these two functions to be determined. The use of isolated cuticles enables cuticular penetration, which constitutes the first step in foliar absorption of chemicals into plant tissues, to be studied. Special consideration will be given to data obtained with the help of short-lived radioisotopes (^{42}K, ^{64}Cu) and a stable isotope (^{10}B). The application of microanalytical techniques for investigating the localization of exogenous elements in treated leaves will be described.

II. DETERMINATION OF THE FACTORS AFFECTING UPTAKE AND TRANSLOCATION

Many results have been reported in the literature concerning the general behavior of inorganic and organic compounds applied to leaves. They have been obtained on entire plants using radioactive isotopes which make it easy to discriminate between the foliar-applied element and the same element already present in the plant or simultaneously taken up by the roots. Unfortunately, radioactive isotopes are not always available or easily usable, thus limiting the investigations to a small number of elements with the well-known radio-isotopes: ^{32}P, ^{45}Ca, ^{59}Fe, ^{65}Zn, ^{35}S, . . . etc. We used the short-lived radioisotopes ^{42}K (T: [half-life] 12.4 hr) and ^{64}Cu (T: 12.8 hr) and an enriched stable isotope (^{10}B) to study foliar absorption of potassium (a macronutrient), and copper and boron (micronutrients).

The experimentation was carried out on plants (lettuce and maize) grown hydroponically in well-defined climatic conditions, the element studied being applied as droplets of solution on the adaxial surface of leaves either directly on the lamina or on a fine nylon textile fabric disk (1 cm^2 in area) held on the leaf by two small strips of adhesive tape. This arrangement made it possible to apply the solution to equal areas of all the leaves treated. At the end of a treatment period which did not exceed 24 hr, being limited by the half-life of the radio-isotope, the treated leaves were detached from the plants. The areas where the droplets had been placed were quickly washed with distilled water so as to recover the residual salt not absorbed by the leaf. Generally, each leaf was given three rinses, and it was shown that the greater part of the radioisotope not taken into the leaf tissue was collected at the first rinse. The adequacy of washing in each experiment was checked by measuring the radioactivity of the three rinse solutions separately, and verifying that the radioactivity of the last one was negligible. When a nylon textile fabric disk was used, it was sufficient to rinse twice and to measure the radioactivity of the disk.

These experiments enabled two main points to be determined: (1) the uptake, as measured by the total radioactivity of the plant after washing of the treated leaf. This was expressed as a percentage of the applied radioactivity or as absolute units. To correct for volumetric error in the dispensing of droplets onto the plant, calculations could be made on the assumption that the radioactivity deposited on each leaf was equal to the sum of the radioactivity of the washing liquids and that of the whole plant. There was no radioisotope leakage into the liquid medium bathing the roots in these experiments. (2) Translocation was estimated by the radioactivity of the untreated parts, and was expressed either as a percentage of the applied radioactivity or as a percentage of the amount taken up.

The influence of various factors on these two different functions can be investigated, the factors that we considered can be classified in two groups: the internal factors connected with the morphology and physiology of the plant, and the external factors related to the characteristics of the solution applied to the leaves. For a specific plant species, the internal factors include the site of the deposit on the lamina, the age of the leaves treated, and the stage of the plant development. There are also factors depending on the root nutritive medium such as the temperature, osmotic potential, and root nutrition affecting the general metabolism of the plant. The external factors include the associated ion in the case of organic salts, the concentration and pH of the solution, the surfactants (wetting agents, etc.), and the interactions with other simultaneously applied chemicals. The influence of factors related to the climatic conditions (relative humidity and temperature) were not considered. The main results obtained from our experimentation on nutrients are presented with a clear distinction between uptake and translocation.

A. Foliar Uptake and Translocation of Potassium

1. Foliar Uptake

The uptake of potassium depends greatly on several characteristics of the applied solution such as the choice of salt and the concentration.[1]

a. Choice of Salt

The uptake of potassium applied as different salts decreases in the following order: K_2CO_3 > KCl-KNO_3-KH_2PO_4 > K_2SO_4. This influence (confirmed for several elements[2-4]) can be better explained by the solubility in water of the salts used and their hygroscopic properties than by the pH of the solutions or the different cellular permeabilities of the anions. Thus, some salts do not completely precipitate after the aqueous phase evaporates, and remain in the form of a saturated solution; continued uptake than depends on the capacity for maintaining the state of a saturated solution. This is related to the solubility and hygroscopicity of the applied compound and to ambient conditions such as the relative humidity.[3-5]

b. Concentration

An increase in the concentration of potassium applied induces a slight decrease in uptake expressed as a percentage of the applied amount, but a large increase in uptake in terms of weight of potassium. This response to concentration declines at higher concentrations (Figure 1) suggesting a progressive saturation of the sites of uptake. Similar trends are observed using other elements.[5-7]

c. Adjuvants

The influence of some wetting agents and of DMSO were tested.

Wetting agents — The addition of a wetting agent is necessary to ensure the adherence of droplets on difficult to wet leaves such as maize leaves. However, not one of the products tested — polyoxyethylene sorbitan monolaurate (Tween® 20), octylphenoxy polyethoxy ethanol (Triton® X100), sodium lauryl sulfate (Empicol® LM), modified sodium dialkyl sulfosuccinate (Montaline® 1304), or dioctyl sulfosuccinic acid sodium salt (Super Montaline® SLT) significantly improved the potassium uptake.[1] The contribution of surfactants is much discussed in the literature, especially in the case of herbicides[1,8-13] but is very complex and still not clarified. Not only must the characteristics of the surfactant be considered but also the specific interactions between plant surface, solute, and surfactant.

DMSO — This substance generated great interest because its capacity to enhance penetration of substances suggested a promising use in the agricultural field.[14]

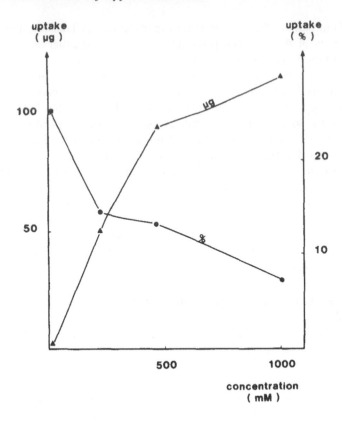

FIGURE 1.   Effect of concentration on the foliar uptake of potassium
as KNO_3 by lettuce over 20 hr.

The results of our experimentation, extended to several elements, revealed that DMSO
can increase the foliar uptake of potassium or iron (Table 1). However, this effect depended
on the salt used. It appeared with the potassium nitrate or chloride but not with the sulfate.

It was found that maize plants treated with foliar applications of KNO_3 with DMSO (1%)
had a higher potassium content and were larger than control plants.[14] DMSO also had a
positive effect on uptake of iron with ferrous sulfate but not with ferric nitrate. There was
no effect on uptake of the phosphate ion, possibly because of the low solubility of phosphates
in DMSO.

The positive effect of DMSO on foliar uptake was partially attributed to its property of
being very hygroscopic; in addition, it depends on the solubility of salts in concentrated
DMSO, their hygroscopicity, and the external conditions (relative humidity and temperature).
So, this effect appears as a purely physical action on the leaf surface. By contrast, DMSO,
at 5 and 10% w/v, inhibited the absorption of potassium and glucose by leaf cells.[16] DMSO
had no effects on ion translocation from leaves. The various results obtained with this
substance constitute an example of the complicated interactions observed with adjuvants.

2. Translocation

It was confirmed that potassium applied to leaves is quickly translocated to young parts
of the plant. However, the treated leaf always contained the major fraction of the amount
taken up, with the greatest concentration in the area previously in direct contact with the
applied solutes (Table 2).

The more important factors for translocation were those related to the plant and to the
nutritive medium.[1] It is clear that translocation depends on the general metabolism of the

Table 1
EFFECT OF DMSO ON THE FOLIAR
UPTAKE OF POTASSIUM AND IRON IN
MAIZE PLANTS OVER 24 HR[14,15]

	DMSO % v/v	Uptake % of the applied radioactivity	% increase
Potassium			
Nitrate	0	31.3	
(10 mM)	0.2	57.6	84
	1	83.7	167
Chloride	0	36.8	
(20 mM)	0.5	98.0	166
	1	92.2	151
Iron			
Sulfate	0	32.7	
(1 mM)	0.5	49.9	53
	1	50.2	54

Note: For each salt the effect of DMSO is significant at 1% level.

Table 2
TYPICAL DISTRIBUTION OF ^{42}K IN A
YOUNG MAIZE PLANT 24 HR AFTER ITS
APPLICATION AS NITRATE, ON THE
SECOND LEAF[1]

	Distribution (%)				
	Treated leaf				
Top	Middle (treated)	Base	Sheath	Aerial parts	Roots
5.1	39.0	11.3	12.2	23.0	9.3

plant as was shown by the influence of root temperature, root nutrition, and the water content of the plant.

a. Age of Leaves

It is well known that the translocation of nutrients or metabolites from leaves depends on their age, thus, potassium was better exported from an old lettuce leaf than a young one.[17] The distribution of potassium in the different parts of the plant was dependent on the position of the donor leaf. The same leaf can both export and import (Table 3).

b. Root Temperature

The translocation of ^{42}K depends to a large extent on the root temperature, it is very low at 4°C, increases rapidly up to 20°C, and then more slowly between 20 and 30°C (Figure 2).

c. Osmotic Potential of the Root Medium

When the osmotic potential of culture solutions was lowered with polyethylene glycol

Table 3
INFLUENCE OF THE AGE OF LEAVES ON THE
DISTRIBUTION OF ^{42}K IN MAIZE PLANTS, 20 hr AFTER
ITS APPLICATION AS POTASSIUM NITRATE

	Distribution (%)							
	Treated leaf		**Other leaves**					
Treated leaf	**Blade**	**Leaf sheath**	**Young**	**Old**	**9th**	**3rd**	**Stem**	**Roots**
3rd	87.3	6.4[a]	0.5	0.8[a]	0.4	—	1.1	3.5[a]
9th	95.7	0.5[a]	0.7	0.2[a]	—	0.1	2.3	0.5[a]

Note: Leaves numbered from base upwards.

[a]   The two values of the column are significantly different at 1% level.

From Chamel, A., *Recent Advances in Plant Nutrition,* Samish, R. M., Ed., Gordon
& Breach, New York, 1971, 395. With permission.

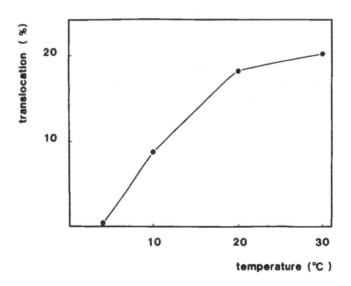

FIGURE 2.   Effect of the root temperature on the translocation of ^{42}K
from a treated leaf in maize plants, over 19 hr.[18]

inducing a water stress in plants, the translocation of ^{42}K from leaves decreased (Figure 3);
the effect was not entirely reversible.[19] This result corroborates the influence of the water
content of plants on the translocation from leaves.

d. Potassium Nutrition

Potassium nutrition of roots is also an important factor affecting the foliar translocation
of this element. The export of ^{42}K from a treated leaf decreased in maize plants grown in
a nutritive solution without potassium prior to the foliar treatment; the effect depended on
the time the roots remained in potassium-free medium (Table 4).

This effect appeared even though these plants had a relatively high potassium content
(more than 2%).

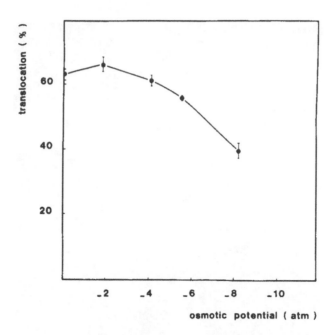

FIGURE 3. Effect of the osmotic potential of the root medium on the translocation of ^{42}K from the treated leaf in maize plants, over $18\frac{1}{2}$ hr. Error bars: standard error. (From Chamel, A. and Bougie, B., *Can. J. Bot.*, 52, 1469, 1974. With permission.)

Table 4

INFLUENCE OF PRIOR POTASSIUM NUTRITION OF THE ROOTS ON TRANSLOCATION OF ^{42}K APPLIED AS NITRATE TO THE THIRD LEAF OF MAIZE PLANTS

	^{42}KNO$_3$ applied at						
	7 days		**14 days**				
Days without potassium	0	7	0	5	9	14	7[a]
% ^{42}K translocated in 22 hr	32.2	9.1	27.3	21.2	13.3	7.0	24.8

Note: Each value is a mean from six replicates. The influence of the factor under consideration is significant at 1% level after 7 and 14 days.

[a] Plants grown 7 days without potassium, then 7 days with potassium.

From Chamel, A., *Recent Advances in Plant Nutrition*, Samish, R. M., Ed., Gordon & Breach, New York, 1971, 395. With permission.

It is clear that foliar translocation depends on the condition of the roots and consequently on the nutritive state of the plant; it can be assumed that foliar translocation could also be affected by other changes in the root zone which would disturb the physiology of the plant.

B. Foliar Uptake and Translocation of Copper

The behavior of copper applied to maize leaves was determined using ^{64}Cu. Uptake is dependent upon several factors related to the characteristics of the liquid deposited on leaves:

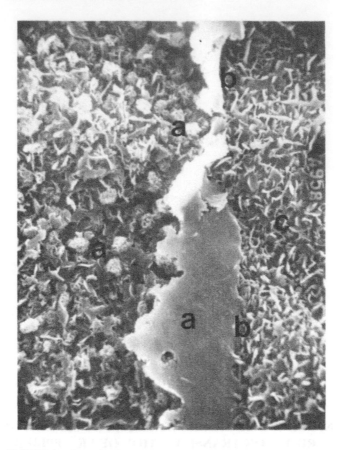

FIGURE 4.   Crystals of cupric acetate on the treated surface after washing.[20] (a) crystallized salt; (b) limit of the deposit; and (c) epicuticular waxes.

it increases with the copper concentration, is significantly lower with sulfate in comparison with nitrate and acetate, and is only slightly affected by the choice of surfactant.[20] The optimum concentration of surfactant was 0.1% with Tween® 20 and Vatsol® OT. Translocation of ^{64}Cu from the treated lamina was limited, however it could be detected in other aerial parts and in the roots.[21] The localization of copper at the site of application (generally more than 90%, 24 hr after the foliar treatment) makes it difficult to determine the degree of cuticular penetration. Several techniques were used to specify the precise localization of the radioisotope. These include:

1.   Observation by scanning electron microscopy of the treated surface revealed that copper acetate crystals were still visible in the epicuticular waxes after the washing operation, following application at a concentration of 0.1 mg Cu per milliliter (Figure 4). This result was concentration dependent. No crystal deposits were detected when sulfate, nitrate, or acetate were applied at concentrations about ten times lower.

2.   The measurement of the radioactivity of the epicuticular waxes extracted by a short immersion (90 sec) of the treated segment in chloroform also revealed the localization on the leaf surface of a fraction of the retained copper. This result observed with the two tested concentrations (6.10^{-3} and 10^{-1} mg Cu per milliliter in the form of cupric acetate), though more pronounced with the higher concentration, was dependent on the rate of evaporation of the aqueous phase of the droplets (Table 5). Thus, washing of the treated leaf was more efficient when the site receiving the deposit was maintained

Table 5
DISTRIBUTION OF COPPER 22 hr AFTER ITS FOLIAR
APPLICATION ON MAIZE PLANTS[21]

% of the applied radioactivity

	Washings		Treated lamina		
	1st	2nd	Epicuticular waxes (e.w.)	Lamina without e.w.	Untreated parts
Zone of the deposit maintained:					
At ambient humidity					
Acetate	86.6	3.6	0.9	8.1	0.7
Nitrate	73.7	10.0	2.9	12.5	1.0
In vapor saturated air					
Acetate	70.5	0.4	0.2	27.8	1.1
Nitrate	59.7	0.5	0.2	38.2	1.5

Note: Copper concentration: $6 \cdot 10^{-3}$ mg Cu/mℓ cupric acetate; each value is a mean from six replicates.

in a high humidity after the foliar treatment, as revealed by the lower values found both in the second washing and in the epicuticular waxes; furthermore, a better penetration was noted in this case. In some cases, sonication was better than washing with HCl to eliminate the nonabsorbed residue.[20]

3. Microanalytical techniques presented in the next part of this chapter were used for the first time to study the localization of exogenous elements in the treated lamina, as a function of the depth from the leaf surface.

The results of these investigations clearly indicate that with elements such as copper, it is necessary to check the efficiency of washing very carefully in order to appreciate the physiological significance of the foliar uptake measurements. The concentration used and the climatic conditions have to be taken into consideration.

C. Distribution of Boron

The case of this element is particularly interesting because it constitutes an example of results obtained with the help of stable isotopes. Indeed, for elements such as boron, the lack of a convenient radioactive isotope makes studies on uptake and translocation very difficult. The distribution of boron applied to leaves as boric acid enriched with ^{10}B (90.5%) was studied using spark-source mass spectrometry[22] for analysis of the stable isotope. This technique was highly sensitive. The results of an experiment reported in Table 6 show clearly that the values of the isotopic ratio ^{11}B/^{10}B in the different parts of the plant differ significantly from the value of the natural abundance (^{11}B/^{10}B: 4.1) thus demonstrating the penetration and translocation of this element. It was calculated that 78 to 93% of boron retained by the plant was still in the treated leaf 24 hr after the foliar application.

This method was recently used to study the penetration and distribution of boron in apple fruit tissues when applied directly to the fruit surface.[50]

D. Comparison with other Elements

The high retention, at the site of application of elements applied to maize leaves was confirmed with ^{65}Zn and ^{59}Fe.[15,21] For example, with zinc applied as 0.1 mM ZnCl$_2$, the fraction present at the site of application represented 81 and 65% of the total radioactivity

Table 6
ABUNDANCE RATIOS OF ISOTOPES (^{11}B/^{10}B) IN
DIFFERENT PARTS OF RADISH PLANTS 24 hr
AFTER A FOLIAR APPLICATION OF [^{10}B]BORIC ACID

Plant part	^{11}B/^{10}B abundance ratios in plants			
	1	2	3	4
Treated leaf	0.8 ± 0.1	0.35 ± 0.05	0.7 ± 0.15	0.3 ± 0.05
Epicotyl	3.3 ± 0.3	2.35 ± 0.25	3.0 ± 0.4	2.2 ± 0.5
Hypocotyl	2.7 ± 0.3	2.0 ± 0.2	1.7 ± 0.25	2.7 ± 0.25

From Chamel, A., Andréani, A. M., and Eloy, J. F., *Plant Physiol.*, 67, 457, 1981. With permission.

of the plant, 1 and 7 days, respectively, after the foliar treatment. The values obtained with iron 24 hr after a foliar application of 1 mM ferrous sulfate or 1 mM ferric nitrate were 98 and 84%, respectively. Zinc was practically never detectable in the epicuticular waxes.

III. LOCALIZATION OF ELEMENTS APPLIED TO LEAVES

A major problem which appeared in the experimentation on whole plants, especially with micronutrients, was to obtain a better understanding of the physiological significance of the fraction recovered in the treated zone which often represented the greatest proportion of the uptake. Two microanalysis techniques were used to determine the exact localization of the exogenous elements in the zone of the lamina corresponding to the site of application. The general principle of these methods is briefly indicated.

A. Laser Probe Mass Spectrography (LPMS)

The apparatus used in this investigation was original equipment built at the Grenoble Nuclear Centre.[23] A light pulse of 10^4 to 10^5 W is focused at a precisely determined point on the specimen to be analyzed. The target tissue undergoes a sudden rise in temperature (10^4 K) and is locally volatilized with the formation of a shallow crater (diameter 3 to 100 μm). The atoms are ionized, accelerated, separated by interaction with a magnetic field, and finally detected on a photographic plate, or more recently, by a high sensitivity detector.[23] After development, the photographic plate shows a number of lines which are characteristic of the different chemical elements and their isotopes. It is possible to perform analysis to greater depths, through the total thickness of the leaf, by repeated laser pulses at the same point, the crater being depressed progressively. The detection limits can be increased by accumulation of spectra from repetitive shots at the same point. New developments in progress enable the performance of the apparatus to be improved as far as the in-depth resolution and sensitivity of the method are concerned.

B. Analysis by Secondary Ion Mass Spectrometry (SIMS)

In this technique,[24] the sample is bombarded by an ion beam. In our investigation, argon ions (5 KeV) were used; the atoms of the sample are emitted as characteristic secondary ions. They are accelerated and sorted by mass spectrometry.

C. Some Applications

The results of several investigations showed that the copper, previously deposited as droplets of cupric acetate on the upper surface of maize leaves, was generally not detectable outside the treated zone and its distribution was heterogeneous inside this area.[21] At all

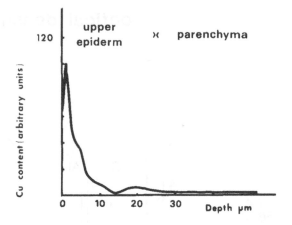

FIGURE 5. Analysis of copper as a function of depth from surface in the treated part of a maize leaf, with a Cameca ionic analyzer, 23 hr after the foliar application.

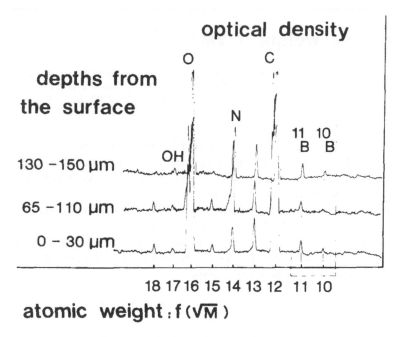

FIGURE 6. Analysis of boron at different depths from surface in the treated part of a radish leaf, with the laser probe mass spectrograph, 24 hr after the foliar application. Thickness of a leaf blade at site of analysis was 150 μm. (From Chamel, A., Andréani, A. M., and Eloy, J. F., *Plant Physiol*, 67, 457, 1981. With permission.)

points analyzed, the copper concentrations rapidly decreased with depth from the leaf surface. The sensitivity of the method did not enable copper to be detected deeper than 20 μm and sometimes much less (thickness of the leaf blade: 0.2 to 0.5 mm) (Figure 5). On the other hand, results with boron revealed that this element applied as boric acid presented a homogeneous distribution with respect to the depth along the analyzed profiles and was not superficially located, as in the results obtained with copper.[22] Figure 6 shows the boron spectra recorded at three depths. It was clear that boron had entered through the leaf surface and was present within the tissues.

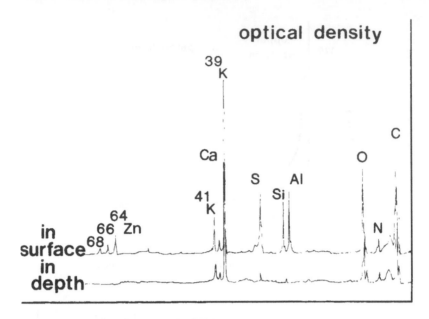

FIGURE 7. Examples of mass spectra obtained with the laser probe mass spectrograph during the analysis of zinc (fungicide: Calyram) in treated part of maize leaf. (From Chamel, A., Marcelle, R. D., and Eloy, J. F., *J. Am. Soc. Hortic. Sci.*, 107, 804, 1982. With permission.)

Laser probe mass spectrography was also used for analyzing zinc applied as fungicides on leaf surfaces.[25] It was shown that zinc was detectable 24 hr after foliar application only in a surface zone to a maximum depth of 30 μm. Two mass spectra corresponding to a surface and an in-depth analysis are given in Figure 7.

These methods are interesting because they enable the distribution of exogenous elements to be specified as a function of depth at the exact place of the deposit. They may be developed providing that the exogenous element can be distinguished from the same element already present in the analyzed tissued, either by using enriched stable isotopes or when the element studied is not detectable in the nontreated tissues.

IV. BEHAVIOR OF CHEMICALS AT THE CUTICULAR LEVEL

The cuticle represents the first barrier to the penetration of exogenous materials deposited on aerial plant surfaces, so cuticular penetration is an essential step to be considered in the analysis of the complex process of foliar absorption.

A. The Cuticle

The plant cuticle is a noncellular, nonliving, lipoidal covering composed of the biopolymer cutin with embedded wax and a layer of epicuticular waxes on the outer surface, the ultrastructure of which varies from amorphous to highly crystalline deposits. Surface morphology appears to be very diversified; the leaf surfaces over vascular tissue often differ in morphology from surfaces over interveinal areas. Furthermore, leaves, and in some cases fruit, have specialized structures namely trichomes and stomata also coated with a cuticle.[26]

Cutin is the major structural framework and is a polyester of hydroxylated fatty acids, most commonly 16 or 18 carbon chain lengths. The epicuticular waxes are composed of

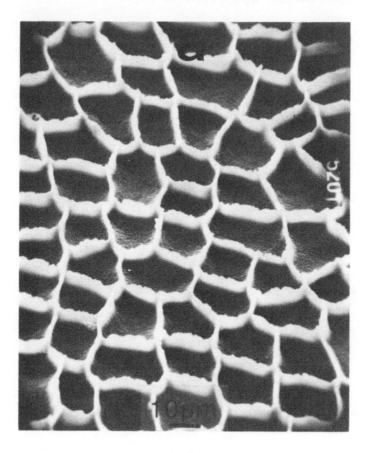

A

FIGURE 8.   Scanning electron micrographs of the internal surface of isolated
cuticles from Passe Crassane pear leaf (A), tomato (B), and Golden Delicious
apple fruit (C).

complex mixtures with long-chain aliphatic and cyclic compounds — primary alcohols (C_{26}, C_{28}, C_{30}), hydrocarbons (C_{29}, C_{31}), secondary alcohols (C_{29}), β diketones (C_{31}, C_{33}), and triterpenoid (ursolic acid) are major constituents. Intracuticular waxes consist principally of fatty acids (C_{16}, C_{18}).[27] The structure of plant cuticles is so heterogeneous that there is no typical plant cuticle and it is impossible to generalize about its morphology and construction.[28]

The wettability and retention of foliar sprays depend greatly on the surface morphology and the nature of chemical groups exposed at the surface.[26] The cuticular behavior of chemicals such as nutritive elements or organic compounds can be studied using isolated cuticles. Cuticular membranes can be separated from the underlying tissues by enzyme mixtures (pectinase and cellulase) or by chemical treatment with ammonium oxalate/oxalic acid or zinc chloride in hydrochloric acid.[29-31] The internal surface of leaf and fruit cuticles, viewed by scanning electron microscopy, is shown in Figure 8. The observation of this surface enables the efficiency of the separation method used to isolate the cuticles from underlying cells to be checked. For penetration experiments, cuticular disks are carefully inspected under a light microscope to select only cuticles without imperfections. The mineral composition of enzymatically isolated cuticles suggests that cuticles have specific ion retention properties.[32]

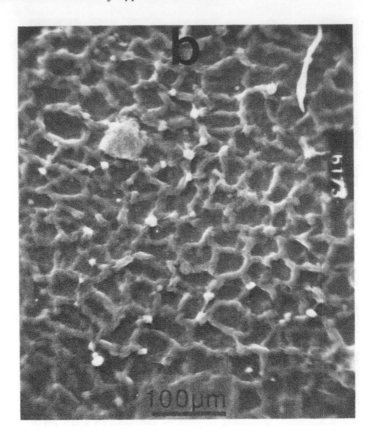

FIGURE 8B

B. Determination of Cuticular Retention and Penetration

The retention of specific chemicals was determined by keeping 10 or 25 cuticular disks (1 cm diameter) in contact with a radioactive solution of the experimental chemical under controlled conditions. After equilibrium was reached, the cuticular radioactivity was measured either as the decrease in radioactivity of the solution or directly on cuticular disks after blotting off with filter paper or quick washing in deionized water to eliminate the excess radioactive liquid adhering to the cuticles. In the first case, the equilibrium was not disturbed so it was possible to calculate the distribution coefficient corresponding to the ratio:[chemical] in the cuticle (dpm/g)/[chemical] in the solution (dpm/g). The relative concentration of the tested compound in the cuticle and its affinity for cuticular material were thus illustrated.

Penetration through isolated cuticles was measured with the following two devices:

1.   A cuticular disk (1 cm diameter) was fixed to the end of a thick-walled glass tube, with the radioactive solution deposited inside in contact with the external surface of the cuticular disk, and with the receiver liquid in a glass scintillation vial on the outside. The liquid levels were equalized to eliminate hydrostatic pressure. After various time intervals, the vials were exchanged and the radioactivity of the whole of the receiver solution was measured each time.
2.   A cuticular disk (1 cm diameter) was inserted between the two symmetrical compartments of a permeability apparatus made of Plexiglas® (Figure 9). The appropriate solutions (10 mℓ) were added to each compartment and stirred by a spinfin magnetic stirbar. The radioactive chemical was added to the outer solution (facing the external surface of the cuticle). At time zero and at various time intervals, a 100-μℓ sample

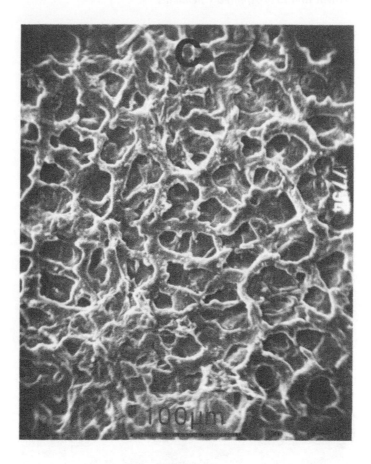

FIGURE 8C

was withdrawn from the inner solution and its radioactivity was measured; 100 μℓ of liquid was immediately added to the receiver after each withdrawal to maintain a constant volume. This more sophisticated device was intended to determine the permeability coefficients well defined by Schönherr[33] and Schönherr and Huber.[34]

Some results of our investigation on transcuticular penetration of nutrients, pollutants, and pesticides are presented below.

1. Calcium[35]

Treatments with calcium salts are frequently applied directly to apple fruit to correct physiological disorders. To improve the efficiency of such treatments, physiological studies on the penetration of calcium into fruit are still necessary. The retention of exogenous calcium at the cuticular level was determined in vitro using chemically isolated apple fruit cuticles (cultivar Golden Delicious). Calcium supplied as $CaCl_2$ can be easily retained by the apple fruit cuticle, more than 80% of this retention occurring within the first minutes, and equilibrium being reached after 1 to 3 hr. It was possible to discriminate between a sorbed fraction corresponding to calcium present as free ion within the intracuticular spaces which was released during washing with deionized water, and an exchangeable fraction attributable to the Ca^{2+} fixed on negative sites of the cuticular matrix, recovered by washing with HCl (Figure 10). The exchangeable fraction could easily be determined due to the high selectivity of cuticular constituents for calcium, which prevented loss of the exchangeable Ca^{2+} during washing (for 180 min) with deionized water.

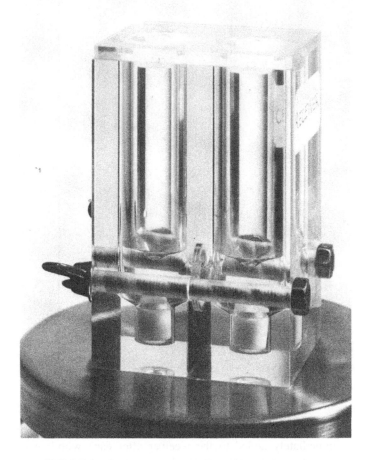

FIGURE 9.   Apparatus used in cuticular permeability studies.

Cuticular retention increased with increasing concentration in the lower range, but it rapidly saturated above 1 meq calcium per liter. It increased greatly with pH increases from 3 to 8 (38.6 to 154.2 μeq calcium per gram for the exchangeable fraction, Figure 11) suggesting an effect of pH on the number of negative charges in the cuticle.

The retention appeared to occur primarily at the cutin matrix level as there was no significant difference between undewaxed and dewaxed cuticles on a surface area basis and greater retention by dewaxed cuticles on a dry weight basis (Table 7). The amount of waxes extracted by chloroform was 370 mg/g of dry cuticular material.[35]

A comparison among several cultivars revealed great variations in retention values which might be related to differences in the chemical composition of the cuticles.

These types of experiments may help to explain the important superficial localization of calcium observed when it is supplied directly to the whole fruit. Recent experiments carried out with the apparatus described above have already clearly demonstrated the influence of pH on diffusion of calcium across isolated apple fruit cuticles. The permeability (self-diffusion) greatly increased with increasing pH.

2. Micronutrients: Copper, Zinc, Manganese

These elements, like calcium, are rapidly retained by enzymatically isolated pear leaf

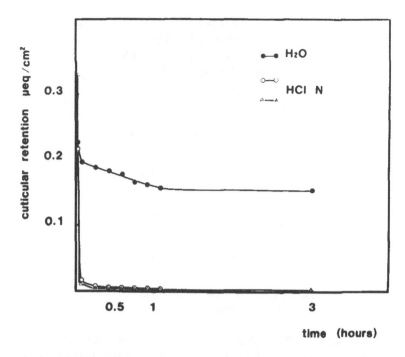

FIGURE 10. Retention of calcium by apple fruit cuticles as a function of washing time with deionized water or HCl. The cuticular disks were initially immersed for 24 hr in unbuffered 10 mM ^{45}CaCl$_2$, pH near 5. (From Chamel, A., *Acta Hortic.*, 138, 23, 1983. With permission.)

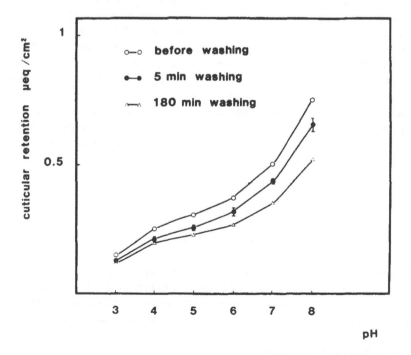

FIGURE 11. Effect of pH on the retention of calcium by apple fruit cuticles. The values obtained before and after washing with water are reported. Ten cuticular disks were immersed for 24 hr in 25 mℓ 10 mM ^{45}CaCl$_2$. Error bars: standard error. (From Chamel, A., *Acta Hortic.*, 138, 23, 1983. With permission.)

Table 7
COMPARISON OF THE
RETENTION OF CALCIUM BY
UNDEWAXED AND
DEWAXED APPLE FRUIT
CUTICLES

	μeq Ca/cm^2	μeq Ca/g
Undewaxed	0.242	71.6[a]
Dewaxed	0.262	124.2[a]

Note: Cuticular disks submitted to 180 min
washing in deionized water after 24 hr
immersion in 10 mM ^{45}CaCl$_2$.

[a] The two values are significantly different
at 1% level.

From Chamel, A., *Acta Hortic.*, 138, 23,
1983. With permission.

cuticles *(Pyrus communis* L. cv Passe Crassane) but this retention is dependent on the element considered (and decreases in the order Cu > Zn > Mn). When cuticular disks initially charged with ^{65}Zn, ^{54}Mn, or ^{64}Cu were submitted to water washing, ^{54}Mn was practically entirely recovered in the washing liquid, whereas only 40% of ^{64}Cu was lost after 48 hr.[36,37] An intermediate value of 80% was noted with ^{65}Zn after a 48 hr wash (Figure 12).

The same retention was observed with cuticles from upper and lower leaf surfaces. However, by floating cuticular disks on radioactive solution it was shown that greater retention occurred from the internal cuticular surface than from the external surface.[36] Several factors such as the concentration and the influence of waxes, already considered in the case of calcium, had similar effects with these elements. The influence of other factors such as the choice of plant species and the date of sampling of leaves was also revealed. Moreover, interactions were demonstrated between divalent ions; thus, copper drastically reduced the cuticular retention of zinc, revealing a high selectivity of copper over zinc.[38] This must be considered in agricultural pratice when several elements are applied simultaneously. As for calcium, pH was an especially important factor since cuticles are polyelectrolytes, their ion exchange capacity depends greatly on pH.[34]

The penetration of these elements through carefully selected pear leaf cuticles separated from the upper astomatous surface, as measured with the glass tube device, was very slow under all conditions. However, the penetration rate was affected by the nature of the receiver, being faster when this consisted of a diluted acid or a salt solution rather than water. Penetration was faster through lower cuticles with stomatal apertures. It must be emphasized that this experimental approach gives values with a high degree of variability, as already reported by several authors.[34,37,39,40] These fluctuations are not related to the experimental conditions but only to the heterogeneity of the cuticular material (despite the careful selection of undamaged cuticles in these studies). Variability could be influenced by differences in the amounts and types of soluble cuticular lipids associated with the cuticles.[41]

3. Example of a Pollutant: Cadmium[42]

The possibilities of cuticular retention and penetration of cadmium were investigated because it is a major environmental contaminant, the above-ground parts of plants receiving cadmium through the atmosphere in areas adjacent to certain metal smelters. It was revealed

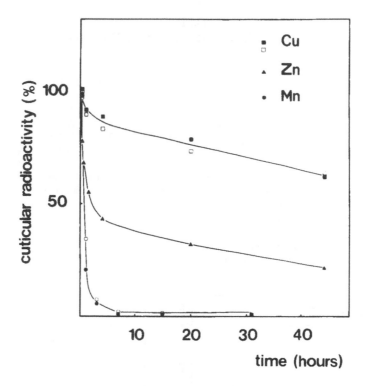

FIGURE 12. Comparative effect of washing with pure water on the cuticular retention of ^{64}Cu, ^{65}Zn, and ^{54}Mn. The cuticular disks were initially immersed for 22 hr in 0.1 mM $^{54}MnCl_2$ or $^{64}CuCl_2$ or $^{65}ZnCl_2$ pH: 4.4. Each value was obtained with ten cuticular disks. (From Chamel, A. and Gambonnet, B., *J. Plant Nutr.*, 5, 153, 1982. With permission.)

that cadmium, as $^{115m}CdCl_2$, may be taken up by isolated cuticles. The mean value calculated from results obtained in the same conditions (ten disks in 25 mℓ solution: 11.8 mg Cd per liter, pH near 6; 5 min washing in deionized water) with cuticles separated from pear leaves, apple and tomato fruit was 1.4 $\pm$ 0.13 μg Cd per square centimeter ($\pm$ standard error). It appeared that the cuticular retention was of the same order of magnitude for leaf or fruit cuticles when expressed as a function of the cuticular surface, but it was greater with leaf cuticles on the dry weight basis. Washing cuticular disks initially immersed in ^{115m}Cd with water or exchange solutions, also enabled the two fractions, sorbed and exchangeable, to be distinguished. The discrimination between these two fractions was possible because the cuticular material presents a relatively good selectivity for this element. The exchange with protons or Zn^{2+} was very efficient. This possibility of exchange is interesting since practical attempts could be made at removing toxic cadmium residues from the aerial epidermal surfaces of contaminated plants.

4. Zinc Applied as Fungicides

The experiments on isolated cuticles clearly demonstrated that the zinc contained in several fungicides was retained at the cuticular barrier level.[25]

5. Herbicides: 2.4 D; 2.4 DB; Dicamba; Glyphosate

The cuticular behavior of several herbicides was investigated more particularly as part of a work relating to the chemical control of the parasitic plant, European mistletoe *Viscum album*.[43] The distribution coefficient was calculated using the indirect method with undewaxed and dewaxed enzymatically isolated mistletoe leaf cuticles. Equilibrium was quickly

Table 8
SOME VALUES OF THE DISTRIBUTION
COEFFICIENT (λ)

Cuticle		2.4 D	Dicamba	2.4 DB[a]
Mistletoe leaf	control[b]	114.4	61.8	215.7
	dewaxed	139.5	69.3	224.7
Tomato fruit	control[b]	98.7	59.2	140.6
	dewaxed	128.6	67.5	170.4

Note: For each herbicide: pH = pKa; concentration = $10^{-3} M$;
solubility (mg/ℓ water, 25°C): 620 (2.4 D) 4500 (Dicamba)
46 (2.4 DB); 25 cuticular disks in 3 mℓ of radioactive
solution for 24 hr.

[a] With 20% ethanol.
[b] Control = undewaxed cuticle.

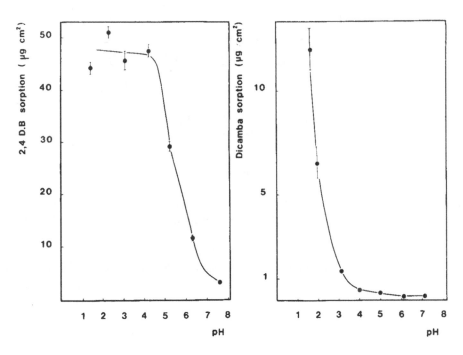

FIGURE 13.   Effect of pH on the sorption of $10^{-3} M$ 2.4 DB and Dicamba by mistletoe leaf
cuticles over 24 hr (error bars: standard error).

reached. The values reported in Table 8, which also relate to tomato fruit cuticles, indicate
that the three herbicides tested are strongly sorbed by the whole cuticle (undewaxed cuticle)
and the cutin matrix (dewaxed cuticle). The variations observed among herbicides also
appeared in the comparison of the radioactivity lost by the cuticular disks during washing
carried out after the sorption period.

Extraction of waxes increased the distribution coefficient by 4 to 30%, so in most cases
the extracted waxes exhibited lower sorption than the cutin matrix. They represented 9.0
and 26.8% of the dry weight of tomato and mistletoe cuticles, respectively. With the weak
organic acid type herbicides, the sorption decreased with increasing pH of the buffer solutions
(Figure 13) and there was a similarity between the effect of this factor both on the sorption

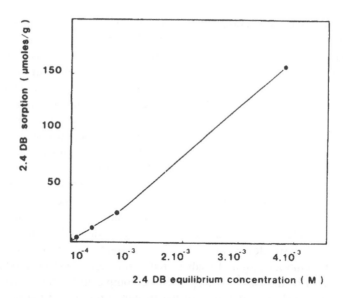

FIGURE 14. Effect of the concentration on the sorption of 2.4 DB by mistletoe leaf cuticles over 24 hr (pH: 4.4).

and on the dissociation of the acid suggesting that it is the nondissociated form which is sorbed by the cuticle.

The sorption of 2.4 DB was directly proportional to the 2.4 DB concentration in the solution (Figure 14), so the distribution coefficient was independent of the herbicide concentration within the range of concentrations tested. Similar results were obtained with naphthaleneacetic acid.[11,33]

The results with glyphosate were very different because they showed that the cuticular sorption of this herbicide is extremely reduced, so it was very difficult to determine the distribution coefficient using only the indirect method. The values obtained for this ratio by measuring directly the radioactivity of mistletoe leaf cuticular disks recovered on filter paper but unwashed were always low, they varied from 1.2 to 8.6 under different pH and concentration conditions.

The cuticular permeability still has to be investigated; it will be interesting to correlate the data obtained with the characteristics of the sorption. The results with herbicides have been limited to pure active substances; however, future investigations should consider the important problem of their formulation.

The problem of the validity of results with isolated cuticles has already been extensively discussed.[10,38,40] A central question is whether studies using isolated cuticles validly represent penetration through the cuticle as it occurs in vivo. The use of isolated cuticles is the only way of studying the specific properties of the plant cuticles in the view of Schönherr and Bukovac[44] and Schönherr and Huber,[34] or of determining the cuticular behavior of chemicals deposited on the aerial parts under natural conditions. However, the penetration in vitro and *in situ* differ as far as the hydration of the cuticle, the evaporation of the applied solution, and especially the nature of the receiver are concerned. It is clear that with an entire leaf, the problem of the receiver (consisting of the external epidermal wall, the plasmalemma, and the cytoplasm) is much more complex. Besides, the separation procedure can affect the cuticular structure.[45-47] The enzymatic method appears to result in cuticles most similar to the in vivo state, and was always chosen in priority. Moreover, cuticular membranes must be isolated and handled at temperatures below 40°C.[48]

The transfer of a substance from the environment to the internal plant tissues comprises several steps involving physical, physiological, and biochemical phenomena. Our analysis of the behavior of exogenous compounds during foliar uptake was limited to the cuticular step. The results of such research can be useful for interpreting those obtained on whole plants. It is clear that the great cuticular affinity for copper demonstrated using isolated cuticles was consistent with the superficial localization of this element in the treated leaves as revealed by microanalytical techniques. Similar observations appeared with zinc supplied as fungicides. For a specific chemical, the characteristics of its cuticular binding and penetration, supplemented when possible with data on the localization of exogenous elements in foliar tissues, can be of assistance in estimating the fraction available for the tissues following a foliar application of this compound. This point represents an important aspect of foliar uptake insufficiently considered in practice. Further investigation is still necessary to find out the possible availability for the plant of the exogenous elements retained at the cuticular level. Moreover, the possibility of directly studying the effect of specific factors such as surfactants on cuticular permeability, with the intention of increasing the penetration of nutrients or herbicides[49], is of great interest for agricultural applications. The improvement of the efficiency of pesticides is an especially important problem to be considered, since any reduction of quantities required leads to reduced residues and lowered costs.

The investigation of the cuticular step must be completed by considering the important surface phenomena such as wettability of the treated surfaces and the participation of the specialized structures: trichomes and stomata, in foliar penetration.[26] The other subsequent steps to take into consideration are cellular absorption through the external epidermal cell wall and the plasmalemma, short-distance transport in the tissues of the treated organ, loading of the conducting vessels, and long-distance transport.

ACKNOWLEDGMENTS

Thanks are due to MM. Barnoud, Carlier, and Fer for helpful discussions.

REFERENCES

1. **Chamel, A.**, Etude à l'aide du ^{42}K de la pénétration et de la migration du potassium fourni par voie foliaire, Thèse de Doctorat d'Etat, Université de Grenoble, France, 1973.
2. **Teubner, F. G., Wittwer, S. H., Long, W. G., and Tukey, H. B.**, Some factors affecting absorption and transport of foliar applied nutrients as revealed by radioactive isotopes, *Mich. Agric. Exp. Stn. Q. Bull.*, 39, 398, 1957.
3. **Koontz, H. and Biddulph, O.**, Factors affecting absorption and translocation of foliar applied phosphorus, *Plant Physiol.*, 32, 463, 1957.
4. **Allen, M.**, The uptake of metallic ions by leaves of apple trees. The influence of certain anions on uptake from magnesium salts, *J. Hortic. Sci.*, 35, 127, 1960.
5. **Barinov, G. V. and Ratner, E. I.**, Some features of the assimilation of substances through the leaves after foliar application, *Fiziol. Rast.*, 6, 333, 1959.
6. **Tukey, H. B., Wittwer, S. H., and Bukovac, M. J.**, Absorption of radionuclides by above ground plant parts and movement within the plant, *J. Agric. Food Chem.*, 9, 106, 1961.
7. **Middleton, L. J. and Sanderson, J.**, The uptake of inorganic ions by plant leaves, *J. Exp. Bot.*, 16, 197, 1965.
8. **Currier, H. B. and Dybing, C. D.**, Foliar penetration of herbicides. Review and present status, *Weeds*, 7, 195, 1959.
9. **Hull, H. M.**, Leaf structure as related to absorption of pesticides and other compounds, *Residue Rev.*, 31, 1970.
10. **Bayer, D. E. and Lumb, J. M.**, Penetration and translocation of herbicides, in *Pesticide Formulations*, Van Valkenburg, W., Ed., Marcel Dekker, New York, 1973, chap. 9.
11. **Bukovac, M. J.**, Herbicide entry into plants, in *Herbicides, Physiology, Biochemistry, Ecology*, Vol. 1, 2nd ed., Audus, L. J., Ed., Academic Press, London, 1976, chap. 11.

12. **Kirkwood, R. C.,** Some criteria determining the efficiency of penetration and translocation of foliage-applied herbicides, in *Herbicides and Fungicides,* McFarlane, N. R., Ed., The Chemical Society, London, 1977, 67.

13. **Price, C. E.,** A review of the factors influencing the penetration of pesticides through plant leaves, in *The Plant Cuticle,* Cutler, D. F., Alvin, K. L., and Price, C. E., Eds., Academic Press, London, 1982, 237.

14. **Chamel, A. and Simiand, J.,** Pénétration et migration d'éléments minéraux appliqués sur les feuilles en présence de diméthylsulfoxyde, *Physiol. Plant.,* 23, 728, 1970.

15. **Chamel, A.,** Pénétration et migration du ^{59}Fe appliqué sur les feuilles de maïs; effet du diméthylsulfoxyde, *Physiol. Plant.,* 26, 170, 1972.

16. **Chamel, A. and Carlier, G.,** Action du diméthylsulfoxyde sur l'absorption du ^{42}K et du glucose-^{14}C(U) par des disques foliares de *Zea mays* et *Pelargonium zonale, Physiol. Plant.,* 32, 128, 1974.

17. **Chamel, A.,** Etude de quelques aspects de l'absorption du potassium par les feuilles à l'aide du radioisotope ^{42}K, *Rev. Potasse,* 3, 31ème suite, 1, 1969.

18. **Chamel, A.,** Effet de quelques facteurs sur la fixation et la migration du ^{42}K appliqué sur les feuilles de maïs, in *Recent Advances in Plant Nutrition,* Samish, R. M., Ed., Gordon & Breach, New York, 1971, 395.

19. **Chamel, A. and Bougie, B.,** Effet du potentiel osmotique du milieu nutritif des racines sur la pénétration et la migration du ^{42}K fourni par voie foliaire, *Can. J. Bot.,* 52, 1469, 1974.

20. **Jourdan, J., Bougie, B., and Chamel, A.,** Etude à l'aide du cuivre 64, de quelques aspects de la rétention du cuivre fourni par voie foliaire, Paper IAEA/Sm-205/26, Proc. Symp. Nuclear Techniques in Animal Production and Health, International Atomic Energy Agency, Vienna, 1976.

21. **Chamel, A. and Gambonnet, B.,** Etude, avec des feuilles *in situ* et des cuticules isolées, du comportement du cuivre et du zinc fournis par voie foliaire, Paper IAEA/Sm-235/21, Proc. Symp. Isotopes and Radiation in Research on Soil-Plant Relationship, International Atomic Energy Agency, Vienna, 1979.

22. **Chamel, A., Andréani, A. M., and Eloy, J. F.,** Distribution of foliar applied boron measured by spark-source mass spectrometry and laser-probe mass spectrography, *Plant Physiol.,* 67, 457, 1981.

23. **Chamel, A. and Eloy, J. F.,** Some applications of the laser probe mass spectrograph in plant biology, *Scanning Electron Microsc.,* 2, 841, 1983.

24. **Paris, N., Chamel, A., Chevalier, S., Fourcy, A., Garrec, J. P., and Lhoste, A. M.,** Applications de quelques techniques nouvelles à l'étude de la microlocalisation des ions dans les tissus végétaux, *Physiol. Veg.,* 16, 17, 1978.

25. **Chamel, A., Marcelle, R. D., and Eloy, J. F.,** Cuticular retention *in vitro* and localization of Zn after a foliar application of zinc-containing fungicides, *J. Am. Soc. Hortic. Sci.,* 107, 804, 1982.

26. **Bukovac, M. J., Rasmussen, H. P., and Shull, V. E.,** The cuticle: surface structure and function, *Scanning Electron Microsc.,* 3, 213, 1981.

27. **Baker, E. A.,** Chemistry and morphology of plant epicuticular waxes, in *The Plant Cuticle,* Cutler, D. F., Alvin, K. L., and Price, C. E., Eds., Academic Press, London, 1982, 139.

28. **Holloway, P. J.,** Structure and histochemistry of plant cuticular membranes: an overview, in *The Plant Cuticle,* Cutler, D. F., Alvin, K. L., and Price, C. E., Eds., Academic Press, London, 1982, 1.

29. **Orgell, W. H.,** The isolation of plant cuticle with pectic enzymes, *Plant Physiol.,* 30, 78, 1955.

30. **Holloway, P. J. and Baker, E. A.,** Isolation of plant cuticles with zinc chloride-hydrochloric acid solution, *Plant Physiol.,* 43, 1878, 1968.

31. **Mazliak, P.,** La cire cuticulaire des pommes (*Pirus malus* L.). Etude morphologique, biochimique et physiologique, Thèse de Doctorat d'Etat, Université de Paris, France, 1963.

32. **Chamel, A., Andréani, A. M., and Mercadier, D.,** Etude par spectrométrie de masse à étincelles de la composition minérale de cuticules isolées de feuilles de Poirier *Passe Crassane, Commun. Soil Sci. Plant Anal.,* 10, 1311, 1979.

33. **Schönherr, J.,** Naphthaleneacetic acid permeability of citrus leaf cuticle, *Biochem. Physiol. Pflanz.,* 170, 309, 1976.

34. **Schönherr, J. and Huber, R.,** Plant cuticles are polyelectrolytes with isoelectric points around three, *Plant Physiol.,* 59, 145, 1977.

35. **Chamel, A.,** Utilization of isolated apple fruit cuticles to study the behaviour of calcium supplied directly to the fruit, *Acta Hortic.,* 138, 23, 1983.

36. **Chamel, A. and Bougie, B.,** Absorption foliaire du cuivre: étude de la fixation et de la pénétration cuticulaires, *Physiol. Veg.,* 15, 679, 1977.

37. **Chamel, A.,** Pénétration du cuivre à travers des cuticules isolées de feuilles de Poirier, *Physiol. Veg.,* 18, 313, 1980.

38. **Chamel, A. and Gambonnet, B.,** Study with isolated cuticles of the behaviour of zinc applied to leaves, *J. Plant Nutr.,* 5, 153, 1982.

39. **Davis, D. G., Mullins, J. S., Stolzenberg, G. E., and Booth, G. D.,** Permeation of organic molecules of widely differing solubilities and of water through isolated cuticles of orange leaves, *Pestic. Sci.,* 10, 19, 1979.

40. **Reed, D. W.**, Ultrastructural Studies on Plant Cuticles: Environmental Effects, Permeability, Electron Microscopy Preparation and Specific Staining, Ph.D. thesis, Cornell University, Ithaca, N.Y., 1979.
41. **Haas, K. and Schönherr, J.**, Composition of soluble cuticular lipids and water permeability of cuticular membranes from citrus leaves, *Planta*, 146, 399, 1979.
42. **Chamel, A., Gambonnet, B., Genova, C., and Jourdain, A.**, Cuticular behaviour of cadmium studied using isolated plant cuticles, *J. Environ. Qual.*, 3, 483, 1984.
43. **Baillon, F., Chamel, A., and Fer, A.**, Comportement du 2,4 DB et du glyphosate marqués appliqués sur le gui; pénétration, transport et redistribution dans le parasite, in compte-rendus de la 12ème Conférence du Columa Tome I, Columa, Paris, 149, 1983.
44. **Schönherr, J. and Bukovac, M. J.**, Ion exchange properties of isolated tomato fruit cuticular membrane: exchange capacity, nature of fixed charges and cation selectivity, *Planta*, 109, 73, 1973.
45. **Norris, R. F. and Bukovac, M. J.**, Structure of the pear leaf cuticle with special reference to cuticular penetration, *Am. J. Bot.*, 55, 975, 1968.
46. **Hoch, H. C.**, Ultrastructural alterations observed in isolated apple leaf cuticles, *Can. J. Bot.*, 53, 2006, 1975.
47. **Jones, J. H.**, Chemical changes in cutin obtained from cuticles isolated by the zinc chloride-hydrochloric acid method, *Plant Physiol.*, 62, 831, 1978.
48. **Schönherr, J.**, Transcuticular movement of xenobiotics, in *Advances in Pesticide Science*, Geissbühler, H., Brooks, G. T., and Kerney, P. C., Eds., Pergamon Press, Oxford, 1979, 392.
49. **Chamel, A.**, Foliar absorption of herbicides: study of the cuticular penetration using isolated cuticles, *Physiol. Vég.*, 24, 491, 1986.
50. **Chamel, A. and Andréani, A. M.**, Demonstration of the penetration of boron in apple fruit using an enriched stable isotope, *Hortic. Sci.*, 20, 907, 1985.

Chapter 3

CHEMICAL REGULATION OF PHOTOSYNTHETIC DECLINE AND LEAF SENESCENCE

Peter M. Neumann

TABLE OF CONTENTS

I. INTRODUCTION

The leaves of plants are designed to utilize water and the energy provided by sunlight in order to convert atmospheric CO_2 and soil-supplied nitrates to the organic nutrients (sugars and amino acids) required for ongoing dry matter accumulation. All of the CO_2 fixed by young developing leaves is retained within the leaf and contributes to leaf growth. As each leaf approaches its final size, the export of organic nutrients, via the phloem, to developing sink tissues in the roots and shoots, is initiated.[1,2] Interestingly, it is just at this stage that photosynthetic capacity in leaves of many species begins to decline progressively.[3-11]

It can be argued that this age-associated decline in photosynthesis, together with associated changes in leaf levels of chlorophyll, protein, and enzyme activities, are part of the preparative stage of leaf senescence.[12,13] This stage is followed by rapid yellowing and death of the leaves, generally denoted as visible senescence. In this chapter the term senescence will be applied to the entire period during which regulated deteriorative changes lead to the eventual death of the leaves.

Crop yields are directly, though not uniquely, related to photosynthesis,[14-16] so that an understanding and eventual control of progressive declines in leaf photosynthesis occurring during senescence are of great theoretical and practical interest. A primary metabolic event accompanying declines in photosynthetic capacity in attached leaves of plants is the decline in the level of the CO_2-fixing enzyme ribulose bis-phosphate carboxylase (RuBP Case).[3-5,9,12,19,20] In addition, declines in noncyclic electron transport[8] and stomatal conductance[9,11,17-19] are observed. Although the course of senescence is being intensively investigated, little is known about the correlative mechanisms regulating the onset of senescence symptoms in attached leaves. It now appears likely that a programmed diversion of nutrients and hormones away from fully expanded leaves towards younger leaves could be one of the control signals involved in the regulation of leaf senescence.[17,21] Reductions in the flux of nutrients and hormones into the leaves of plants can also be caused by a reduction in the supply capacity of the root-soil complex. In either case, it might be possible to increase the flux of a limiting solute into leaf cells by foliar spray applications, and thus to inhibit declines in the photosynthesis.

This chapter will review (1) evidence linking developmental changes in the flux of endogenous nutrients and phytohormones with leaf photosynthetic capacity and senescence and (2) the promotive and inhibitory effects of foliar applications of exogenous chemicals on leaf photosynthesis and senescence.

II. SOLUTE FLUX AND LEAF METABOLISM

The net flux of endogenous solutes (amino acids, mineral ions, and phytohormones) and water into leaf cells is related to the relative rates of xylem and phloem transport in and out of the leaf. Mass flow of xylem solution, caused by the water potential differences between the soil-root complex, leaves, and atmosphere, is the primary factor driving the initial supply of xylem solutes from the roots to the leaves. Distribution of neutral and negatively charged solutes, between competing leaves of rapidly transpiring plants, is largely a function of xylem pathway and stomatal resistances.[22,23] Stomatal resistances often increase progressively during leaf aging (Tables 1 and 2 and References 9, 11, 17-19, 23, 24) so that older leaves receive lower inward fluxes of xylem solution and atmospheric CO_2 than young leaves with fully open stomata. Table 3 lists numerous reports showing that induced reductions in xylem flux of nutrients or growth regulators into leaves can have wide ranging and rapid short-term effects on leaf metabolism, including photosynthesis. Moreover, when the flux of mineral ions (nitrogen, phosphorus, and potassium) into leaf cells is reduced by cultivation on nutrient-deficient root media, adverse long-term effects on photosynthesis are clearly

Table 1
CHANGES IN COMBINED (ABAXIAL + ADAXIAL) STOMATAL RESISTANCES IN LEAVES OF BEAN (*PHASEOLUS VULGARIS* L.) FROM EARLY VEGETATIVE TO REPRODUCTIVE GROWTH (MEAN ± SE, n = 6)

Leaf	2 weeks	3 weeks	4 weeks	5 weeks
Primary (base)	2.9 ± 0.4	4.7 ± 0.6	6.3 ± 0.7	7.0 ± 0.4
1st Trifoliate	—	2.9 ± 0.2	4.7 ± 0.6	6.7 ± 0.6
2nd Trifoliate	—	2.7 ± 0.2	4.2 ± 0.3	6.1 ± 0.5
3rd Trifoliate	—	—	2.6 ± 0.2	5.7 ± 0.8

Adapted from Neumann, P. M. and Stein, Z., *Physiol. Plant.*, 62, 390, 1984.

Table 2
DISTRIBUTION OF WATER LOSS RATES BETWEEN DIFFERENT LEAVES ALONG STEM OF COTTONWOOD SEEDLINGS

Leaf number based on plastochron index	Water loss (mℓ/100 cm^2/hr)
0—5 (young)	2.57
6—8	2.70
9—13	2.92
14—18	2.07
19—23	1.57

Based on information obtained through personal communication, R. E. Dickson.

Table 3
EFFECTS OF ALTERED RATES OF DELIVERY OF INDICATED XYLEM SOLUTES ON LEAF TISSUE METABOLISM

Xylem solute	Response	Response time	Tissue	Ref.
PO_4	Photosynthesis	20 min	Spinach	25
NO_3	NO_3 reductase level	5 hr	Corn	26
Zeatin	Stomatal resistance	48 hr	Soybean	27
BA	Free amino acids	6 hr	Oats	28
B	Nyctinastic movement	20 min	Albizzia	29

observed.[4,5,30-33] A particularly close correlation was found between leaf nitrogen and potassium levels, RuBP Case, and CO_2 exchange capacity during the expansive growth of rice leaves (Table 4). However, in a subsequent study using ^{15}N labeling it was shown that relatively high levels of nitrogen in the root media did not prevent degradation of RuBP Case and net export of nitrogen from fully expanded mature leaves.[4] thus, while photosynthetic capacity in young expanding leaves was responsive to nitrogen supply from the root media, a high nitrogen supply potential could not prevent programmed declines in photosynthesis during leaf aging.

Table 4
CORRELATION (r) BETWEEN CONTENTS OF
NITROGEN, PHOSPHORUS, AND POTASSIUM IN
EXPANDING 12TH LEAF BLADE OF
HYDROPONICALLY GROWN RICE PLANTS OVER A
13-DAY PERIOD AND LEVELS OF PHOTOSYNTHESIS
OF RuBP CASE

	Nitrogen	Phosphorus	Potassium
Photosynthesis ($\mu\ell$ CO_2 min^{-1} leaf blade^{-1})	0.984	0.518	0.802
RuBP Case (mg leaf blade^{-1})	0.981	0.451	0.794

Adapted from Makino, A., Mae, T., and Ohiro, K., *Soil Sci. Plant Nutr.*, 30, 63, 1984.

Table 5
RELATIVE AVAILABILITY OF
SOLUTES FOR IMPORT INTO
LEAF TISSUES VIA PHLOEM
TRANSPORT[1,2,6,21,22,34-36]

Young leaf	Low	Medium	High
	Ca	Mn	N
	B	Zn	P
	NO₃	Fe	K
		Cu	S
		Mo	Mg
			Cl
			Ni

Mature Leaf: low for all solutes

Fluxes of nitrogen and other essential solutes in and out of leaves are also dependent on the amount and direction of transport in the second whole plant transport system, the phloem. The data in Table 5 indicate that the relative mobility of different solutes in the phloem transport system can vary considerably and that the phloem system transports solutes out of — but not into — mature leaves. The inevitable conclusions are (1) developing leaves are totally dependent on the xylem only for incoming supplies of calcium, boron, and nitrate, which are relatively phloem immobile and (2) mature leaves are totally dependent on the xylem for all supplies of inorganic ions, amino acids, cytokinins, and water provided by the roots.

The dependence of mature leaves on xylem supplies, together with the rapid effects of induced changes in xylem supply rate on leaf metabolism (Table 3) suggest that studies of ontogenetic changes in net flux of solutes essential to leaves could throw some light on the control mechanisms involved in leaf development.

Pate and Atkins[6] followed the changes in import and export of nutrients during the course of development of a single leaf attached to a lupin plant. Some of their results are shown in Table 6. The leaves reached maximum area and dry weight at around day 20 and thereafter there was a net efflux of nutrients (nitrogen in this case). Interestingly, both CO_2 uptake and chlorophyll levels also began to decline after 20 days, suggesting a possible correlation

Table 6
RELATIVE IMPORT (+) AND EXPORT (−) OF NITROGEN (MG PER INDICATED NUMBER OF DAYS) BY UPPERMOST LEAF OF *LUPINUS ALBUS* L. BETWEEN 1 AND 66 DAYS

	1—11 days	11—20 days	20—38 days	38—66 days
Xylem	+1.85	+5.83	+3.75	+6.53
Phloem	+0.42	−1.07	−4.58	−8.98
Net change	+2.27	+4.76	−0.83	−2.45

Adapted from Pate, J. S. and Atkins, C. A., *Plant Physiol.*, 71, 835, 1983.

Table 7
RATES OF ALTERATION IN PHYSIOLOGICAL PARAMETERS ASSOCIATED WITH LEAF PHOTOSYNTHETIC PRODUCTIVITY DURING AGING OF ATTACHED PRIMARY LEAVES OF BARLEY

Leaf parameter	Assay period, days after planting	Time for doubling or halving initial value
Transpiration (−)	12—22	3.0
Stomatal resistance (+)	12—22	3.0
RuBP Case (−)	10—22	5.2
Photosynthesis (−)	12—22	5.4
Mesophyll resistance (+)	12—22	5.5
Chlorophyll (−)	14—22	10.4

Note: Signs indicate whether parameter increased (+) or decreased (−) with age.

Adapted from Friedrich, J. W. and Huffaker, R. C., *Plant Physiol.*, 65, 1103, 1980.

between the onset of net nutrient efflux and declines in photosynthesis. Other reports support the idea of a correlation between declining transpirational flux into expanded leaves of various plant species and declining photosynthetic capacity. Thus, Crafts-Brandner et al.[37] concluded that the flux of xylem-supplied NO_3 into leaves of three varieties of maize plants was a factor regulating rates of senescence development. The data of Catsky et al.[11] reveal that declines in stomatal conductivity and transpiration rates preceded (by 3 days) the onset of progressive declines in photosynthesis (net CO_2 influx) in mature primary leaves of bean plants. Similarly, Friedrich and Huffaker[9] found that increases in stomatal resistance and declines in transpiration rates accompanied senescence-related declines in RuBP Case level, RuBP Case activity, and photosynthesis in attached primary leaves of barley seedlings (Table 7). Friedrich and Huffaker[9] concluded that RuBP Case enzyme activity most closely correlated with photosynthetic activity during leaf senescence. However, they also showed that the most rapid rate changes occurring during leaf aging were declines in transpiration and stomatal resistance and that these exceeded rates of decline of photosynthesis over a 10-day period (Table 7). In conclusion, these data suggest that changes in solute flux are closely associated with the declines in photosynthetic activity occurring during the aging and visible senescence of attached leaves.

Table 8
SYMPTOMS OF REJUVENATION
INDUCED IN ATTACHED PRIMARY
LEAVES OF COMMON BEAN BY
DETOPPING

	Intact	Detopped
Photosynthesis, $\mu\ell$ O_2 dm^{-2} min^{-1}	90[a]	120[b]
Soluble protein, mg g fresh wt^{-1}	13[a]	14[a]
Chlorophyll, mg dm^{-2}	4.3[a]	5.6[b]
Primary leaf fresh wt (g)	1.47[a]	2.00[b]
Stomatal resistance, s cm^{-1}	7.5[a]	5.5[b]
Xylem solute flux,[A] μg acid fuchsin dm^{-2} min^{-1}	18[a]	33[b]

Note: Detopping (removal of shoot above primary node) was
at 3 weeks and leaves were assayed at 4 weeks. Silver,
Stein, and Neumann, unpublished results. Means of 6—
30 assays. Data followed by different letters differ significantly ($p = 0.05$).

[A]   Excised transpiring shoots assayed after imbibing 0.5%
w/v acid fuchsin solution for 15 min.

III. REVERSING LEAF SENESCENCE

It has been argued in the previous sections that declines in the photosynthetic capacity of
expanded leaves may be associated with declines in the net flux of essential solutes into the
leaf cells. If this is so, then maintaining or increasing the net flux of essential solutes into
leaf cells should prevent or reverse leaf senescence symptoms such as progressive declines
in photosynthetic capacity. Several lines of evidence suggest that this may indeed by the
case.

A. Defoliation

Surgical removal of vegetative or reproductive sink tissues can have rapid effects on levels
of photosynthesis and senescence of the remaining leaves in many plant species. Whereas
removal of reproductive organs can either accelerate[37] or delay[13,19,38] leaf senescence symptoms, it is generally observed that partial defoliation leads to retardation of senescence and
increases in photosynthetic activity of the remaining leaves.[8,39-44] Typical results of a defoliation experiment with bean plants (Table 8) show that higher rates of photosynthesis and
other signs of leaf rejuvenation were observable within 7 days of detopping.

Wareing et al.[44] suggested that partial defoliation reduced competition among remaining
leaves for mineral nutrients and/or cytokinin phytohormones supplied by the roots, thus
causing the increase in photosynthetic efficiency. However, Neumann and Stein[39] recently
reported that detopping rapidly reduced mineral ion transfer out of the roots of bean plants
and concluded that increased flux of mineral ions into the remaining leaves of detopped
plants was unlikely to be the cause of their rejuvenation. They concluded that influx of
cytokinins into leaves was probably increased and/or that the remaining leaves of detopped
plants respond to less negative xylem water potentials and increased cell turgor. These two
mechanisms can be related since cytokinin accumulation is reduced by water stress.[45,46]
Increased levels of cytokinin activity in leaves of detopped beans and other plants were only
detected 8 and 20 days after detopping,[47,48] i.e., following the onset of increased photosynthetic activity between 2 to 7 days.[8,39-41] However, cytokinin activity in roots and stems of

Table 9

CHLOROPHYLL AND SOLUBLE PROTEIN LEVELS IN LEAVES OF INTACT PLANTS OR EXPLANTS WITH ONE POD PER NODE EXCISED AT MID-PODFILL AND MAINTAINED ON COMPLETE MINERAL SOLUTION WITH ZEATIN (4.6 μM)

Treatment	Chlorophyll (mg/g fresh wt)	Soluble protein (mg/g fresh wt)
Intact Plants		
Initial	4.3 ± 0.1	16 ± 0.5
Final	0.6 ± 0.2	5 ± 1.0
Explants		
Mineral solution + zeatin	2.5 ± 0.3	12 ± 1.0
Zeatin	0.8 ± 0.3	11 ± 1.5
Mineral solution	0.8 ± 0.1	8 ± 1.0
H_2O	0.4 ± 0.2	6 ± 1.0

Note: Green leaves were assayed initially at midpodfill. Explants on H_2O or mineral solution showed accelerated onset of yellowing and were assayed after 13—18 days; other treatments were assayed when intact plants initiated yellowing after 27 days. Mean ± SE, n = 4.

Adapted from Neumann, P. M. and Noodén, L. D., *J. Plant Nutr.*, 6, 735, 1983.

detopped plants increased within 2 days of detopping.[47] Moreover, Palmer et al.[49] found an increase in levels of dihydrozeatin-O-glucoside, a cytokinin metabolite, in primary leaves of bean plants within 5 days of detopping, suggesting that excess-free cytokinins may have been rapidly metabolized to this less active form. Defoliation studies therefore lend support to the hypothesis that increases in the flux of a limiting endogenous solute (cytokinins in this case) into mature leaves can be associated with changes in development. More direct evidence comes from experiments assaying the effects of exogenous chemicals (mineral ions and cytokinins), on photosynthesis and senescence in expanded mature leaves.

B. Exogenous Chemicals
1. Cytokinins and Macronutrients

Neumann and Noodén[27] investigated the relative contribution of mineral ion supply and cytokinin in the regulation of monocarpic senescence in soybeans. This type of leaf senescence is associated with the filling of the pods and is characterized by a rapid onset of yellowing and leaf abscission. Instead of using whole plants, the course of senescence was assayed in excised explants consisting of an expanded trifoliate leaf, subtending pods and a 10-cm stem section, the cut base of which was incubated in water or nutrient solution. When the podded explants were incubated in water alone, the leaves initiated accelerated senescence (by comparison with intact plants). Incubation in a complete mineral solution delayed by only a few days the accelerated onset of leaf senescence. When the explants were incubated in a cytokinin solution (4.6 μM zeatin) leaves turned pale green, but yellowing and abscission were prevented.[27] A combination of mineral nutrients with cytokinin induced a dark green coloration in the leaves and completely prevented their abscission even though the pods turned from green to brown during the same period.[50] Table 9 shows that levels of chlorophyll and soluble protein (as senescence indicators) in leaves of explants on cytokinin plus minerals exceeded those in leaves of explants on other solutions or of equivalent leaves left on whole plants, 27 days after initiation of the experiment. Seed weights of explants

Table 10
EFFECTS OF FOLIAR SPRAYS OF MINERAL NUTRIENTS (NITROGEN, PHOSPHORUS, POTASSIUM, AND SULFUR) AND/OR PLANT GROWTH REGULATORY SUBSTANCES ON INDICATORS OF LEAF SENESCENCE AND PLANT GROWTH

	Spray treatment					
Assay	Control	NPKS	BA	BA + NPKS	BA + auxin	BA + auxin + NPKS
Soluble protein (mg/g fresh wt PRI-LF)	13[a]	15[b]	17[b]	13[a]	14[a]	13[a]
Chlorophyll (mg/dm^{-2} PRI-LF)	4.3[a]	4.3[a]	5.1[b]	4.8[a]	5.0[b]	4.9[b]
Fresh wt (g/PRI-LF)	1.47[a]	1.53[b]	1.90[b]	1.42[a]	2.00[b]	2.35[b]
Root dry wt (g)	0.23[a]	0.25[b]	0.17[b]	0.21[a]	0.18[a]	0.18[b]
Shoot dry wt (g)	0.58[a]	0.65[b]	0.62[b]	0.53[a]	—	—

Note: Foliar sprays were applied to run off to the primary leaves (PRI-LF) of 3-week plants taking care to avoid contamination of root zone. Spray solutions; NPKS = 0.025 M KNO$_3$, 0.016 M NH$_4$H$_2$PO$_4$, and 0.0016 M K$_2$SO$_4$; cytokinin = 20 ppm benzyladenine (BA), auxin = 3.75 ppm naphthalene acetic acid. Spray solutions made up with 0.1% w/v glycerine and Tween® 80 to aid wetting and penetration. Results are means of separate assays of 8—30 plants 1 week after spray treatment. Shoot dry weight refers to material above the primary node. Results with the same superscript letter are not significantly different (p = 0.05). Silver, S., Stein, Z., and Neumann, P. M., unpublished results.

on mineral nutrients with cytokinin increased during the experiment and final dry weights were equivalent to those in intact plants with one pod per node.[50] This suggests that leaf photosynthetic activity and carbon compound transfer from leaves to the seeds were responsive to an ongoing flux of minerals and cytokinins into the leaf.

The above results raised the possibility that a combination of mineral nutrients together with a cytokinin, applied as a foliar spray, might prove to be more effective than application of either one alone. Foliar applications of cytokinins have been shown to retard senescence in the attached leaves of many species including bean, oat, soybean, and wheat.[51-56]

With regard to minerals it has been argued that late-season foliar applications of macroelements nitrogen, phosphorus, potassium, and sulfur (NPKS) may enhance yields by preventing loss of leaf photosynthetic efficiency associated with mineral nutrient drain and leaf senescence during the seed-filling period of plant development.[57,58] However, a considerable body of information suggests that late-season foliar applications of nutrients, if and when they do increase yields, do not do so by preventing leaf senescence symptoms.[59] Thus, when leaf senescence in soybeans was prevented by genetic or hormonal manipulation, no significant yield increases were observed in either growth chamber or field trials.[52,60,61] Despite these reservations, it seemed of interest to determine the effects of combinations of mineral ions and cytokinins on the metabolism of the expanded primary leaf of bean plants. The ontogeny of this leaf has been extensively characterized in my laboratory[17,39] and it shows characteristic senescence-associated declines in chlorophyll, protein, and photosynthesis after full expansion at 3 weeks. The results of experiments designed to test the effect of a single foliar application of mineral nutrients and/or synthetic phytohormones on subsequent development of a 3-week leaf are shown in Table 10. A single foliar spray of NPKS applied to the primary leaf alone slightly stimulated the ongoing growth of roots, shoots, and the treated leaf within 1 week. Treatment with naphthalene acetic acid (NAA) and/or benzyladenine (BA) tended to dramatically increase growth and alter the appearance (to dark green

Table 11
**EFFECTS OF IRON DEFICIENCIES ON GROWTH OF
BEAN SEEDLINGS**

Treatment	Length of Roots (cm)	Roots (g fresh wt)	Tops (g fresh wt)
+Fe[a]	31.0 ± 4.6	2.0 ± 0.3	6.7 ± 0.5
−Fe[a]	14.2 ± 1.1	0.9 ± 0.5	4.4 ± 0.7
−Fe,FeSO₄ spray	29.6 ± 1.6	2.3 ± 0.6	7.5 ± 1.2
−Fe,FeEDDHA spray	18.6 ± 3.9	1.2 ± 0.3	4.8 ± 0.4

Note: Seedlings (7 days old) were sprayed to runoff with solutions containing 800 $\mu g/m\ell$ iron (as $FeSO_4$) or 200 $\mu g/m\ell$ iron (as Fe EDDHA) and 0.04% (w/v) L77 surfactant. The plants were harvested at 19 days.

[a] ± Fe indicates continual presence or absence of chelated Fe in root media

Adapted from Neumann, P. M. and Prinz, R., *Plant Physiol.*, 55, 988, 1975.

and leathery) of the treated primary leaves. In fact, treatment with NAA/BA resulted in primary leaves, similar to those obtained as a result of the defoliation treatment (Table 8). However, growth of the roots and shoot tended to be reduced by hormone applications, suggesting that these treatments may have turned the treated primary leaves into strong competitive sinks for photosynthate, at the expense of other plant parts. The combination of minerals with hormones did not result in any dramatic synergistic effects on the parameters assayed, by comparison with hormones alone. These results reveal clearly that the sequential senescence of mature leaves on bean plants can be controlled by increasing the flux of auxin and cytokinin via foliar application. Auxin-cytokinin combinations also prevented mono-carpic leaf senescence in soybeans.[52] Whether such applications can positively affect photosynthesis and yield characteristics of beans or other species in the field remains to be determined.

2. Micronutrients

When the soil-root complex provides a less than adequate flux of plant-essential micro-nutrients (such as iron, manganese, and zinc) to the leaves, the leaves respond with pro-gressive losses in functional capacity and eventual death. Accelerated leaf senescence and associated loss of photosynthetic capacity caused by micronutrient deficiency can often be effectively reversed by foliar nutrient applications. Indeed, such foliar sprays are widely used to relieve micronutrient deficiency in commercial practice.[62,63] In this respect, the plant-essential microelements are similar to plant growth regulatory substances (e.g., cytokinins), i.e., very small increases in flux into the cells, attainable with a single foliar application at low concentrations, can have dramatic effects on cell and organ (leaf) development. The results of some experiments on the relief of iron deficiency in seedlings[64] illustrate this point. Table 11 shows that a single foliar spray of FeSO₄ at 800 $\mu g\ m\ell^{-1}$ maintained the growth (weight increases) of roots and shoots of 7-day bean seedlings on iron-deficient media over a subsequent 12-day period at levels equivalent to those attained by plants grown continuously on iron-containing root media, i.e., leaf photosynthetic capacity was maintained. The pos-sibility of using foliar sprays of micronutrients, such as iron, to potentiate the early estab-lishment of an effective shoot and root system by young seedlings growing in deficient (e.g., calcareous) soil conditions in the field needs to be investigated. Clearly, the overall effec-tiveness of spray treatments depends on the phloem mobility of the applied micronutrients out of the treated leaf to other developing leaves and to the roots. Table 12 shows that a single foliar application of iron enabled the roots to maintain a high level of metabolic

Table 12
ACTIVITY OF ROOTS
MEASURED BY TIME TAKEN
FOR REDUCTION OF
TRIPHENYL TETRAZOLIUM
CHLORIDE

Treatment[a]	Time (min)
+Fe	23 ± 6
−Fe	∞
−Fe,FeSO$_4$ spray	15 ± 9
−Fe,FeEDDHA spray	60 ± 9

Note: Roots were examined every 5 min until red
color appeared in root tips. Results ± S.D.
are the average of observations on five roots
from each treatment.

[a] Treatments as in Table 11.

Adapted from Neumann, P. M. and Prinz, R.,
Plant Physiol., 55, 988, 1975.

activity, as shown by rapid reduction of triphenyl tetrazolium chloride (TTC), even after 12 days on iron-free root media. Since great care was taken to avoid contamination this suggests that iron was effectively transferred out of the treated leaf to the developing roots of the seedlings via the phloem.

A more direct assay of the relative mobility of foliar-applied nutrients can be obtained by using radioisotopes. Neumann and Chamel[35] recently investigated the mobility of the putative plant micronutrient, nickel,[65-67] by comparing transport of applied ^{63}Ni out of treated leaves to root and shoot sink tissues of pea seedlings with that of ^{86}Rb (potassium) and ^{45}Ca. ^{63}Ni was found to be transported, via the phloem, as effectively as ^{86}Rb, whereas ^{45}Ca was, as expected, relatively immobile.

3. Carbon Dioxide

Elevated levels of atmospheric CO_2 can significantly increase leaf photosynthesis (Chapter 9). In this section, some recent reports touching on the relationships between elevated levels of CO_2 and onotogenetic declines in photosynthesis during leaf senescence in CO_2-fertilized field crops will be considered.

Havelka et al.[68] applied high CO_2 levels (1200 $\mu\ell/\ell$) to winter-grown wheat plants, enclosed in open-top polyester chambers, at three different stages of growth. They found that a 17% increase in seed yield was attainable when plants were exposed to high CO_2 during the period between jointing and anthesis (i.e., before seed set and filling).

Interestingly, in the present context, high CO_2 had no significant effect on flag leaf chlorophyll and protein content or the onset of their decline during flag leaf senescence. Similarly, no effects on senescence-induced declines in RuBP Case were observed. However, elevated CO_2 levels did result in a 50% increase in apparent rates of photosynthetic fixation of $^{14}CO_2$ by the flag leaves. Since CO_2 levels did not retard senescence, the stimulation of photosynthesis was stated to be a result of the well-known inhibitory effect of high CO_2 on the oxygenase activity of RuBP Case which effectively converts fixed carbon back to CO_2. By contrast, attempts at directly inhibiting the oxygenase activity of this enzyme in vivo with foliar sprays of regulatory chemicals have not yet had much success.[69] Elevated CO_2

also increased accumulation of starch and sucrose in the flag leaf. The levels of stored carbohydrates in the flag leaf were rapidly depleted during subsequent seed growth suggesting that in wheat, seed yield was a function of high carbohydrate availability during the period in which seed number was being determined.

Similar CO_2 fertilization trials were conducted with soybeans of determinate and non-determinate varieties. Nondeterminate plants were found to be more responsive to CO_2 enrichment, giving yield increases between 56 and 81% by comparison with maximum yield increases of 36% with a determinate variety.[70,71] By contrast with wheat, significant increases in soybean yield were obtained only when CO_2 enrichment was continuous from emergence to maturity or from early pod fill to maturity.[71,72] Apparently, high CO_2 levels during the seed filling period in soybean resulted in less pod or seed abortion, since the increased yield was associated with greater numbers of seeds rather than increased weight per seed. CO_2 enrichment did not prevent senescence-associated declines in photosynthesis, or in levels of leaf chlorophyll and protein. However, levels of photosynthesis in CO_2 treatments were higher than for control plants until late into the growing period. Elevated levels of photosynthesis in CO_2-enriched soybean plants were maintained despite CO_2-induced lowering of leaf stomatal conductances, suggesting that water use efficiency (grams H_2O lost per gram CO_2 fixed) must have been improved. Rogers et al.[73] showed directly that CO_2 enrichment significantly enhanced the ability of soybean leaves to withstand water stress treatments; this may also contribute to the yield enhancing effects of CO_2 fertilization.

IV. EXOGENOUS CHEMICALS AND DAMAGE

Foliar application of various plant regulatory chemicals has been practiced for over 100 years.[74] One of the main limitations associated with this mode of application is the induction, by excess levels of applied chemical, of necrotic burn damage at the site of application. Necrotic damage can adversely affect leaf photosynthesis in several ways:

1. There is a direct loss of photosynthetic tissue and hence capacity.
2. Export of photosynthate to developing sink tissues can be directly inhibited as a result of damage to the phloem.
3. Indirect damage responses in the leaf, e.g., changes in hormone or protein levels may affect photosynthetic efficiency.
4. The desire to avoid damage often places limits on the composition and concentration of chemicals used for leaf application and hence on the amount and effectiveness of the response which can be expected.

Despite the importance of the damage potentially caused by leaf-applied chemicals, relatively little attention has been paid to development of a better understanding, and thus control, of the underlying mechanisms involved; (e.g., two recent reviews on foliar nutrition[62,63] do not index the subject of leaf damage.)

An understanding of the mechanisms of damage induction by leaf-applied chemicals is also relevant to numerous ecological problems associated with salt sprays, acid rain, and atmospheric pollutants such as ozone, sulfur dioxide fluorides, nitric oxides, and ethylene. In the following sections possible mechanisms involved in induction of leaf damage by various leaf-applied chemicals will be discussed.

A. Surfactants

Surfactants are added to foliar spray solutions in order to increase surface wetting and, at sufficiently low surface tensions, stomatal penetration.[75] However, many surfactants can be phytotoxic and cause burn damage in their own right. Neumann and Prinz[75] reasoned that the relative toxicity of surfactants could be related to the damaging effects of the surfactants on plant cell membranes. They developed a rapid and quantitative bioassay based on determining spectrophotometrically the leakage of anthocyanin pigment from cells of uniform beetroot discs damaged by exposure to surfactant solutions at 0.01 to 0.10% w/v. Several surfactants were found to cause irreversible damage to beet cell membranes. The most damaging surfactant tested, Aerosol® OT, also caused serious damage to citrus trees in field trials.[76] Some surfactants which caused little or no damage (L77 and Tween® 80) are widely used in agricultural practice.

B. Mineral Nutrient Solutions
1. pH
Reed and Tukey[76] investigated the effects of pH on foliar absorption of phosphorus compounds applied as droplets and left for 48 hr on intact chrysanthemum leaves. The most dramatic effect was an increase in uptake at pH 2 and this was associated with the induction of necrotic damage at the site of application. At pHs above 2, the main effects on penetration seemed to be via the relative solubility, moisture retention, and degree of crystallization of salts on the leaf surface, but no damage was noted.

2. Osmotic Damage
Some of the damage caused to leaves by exposure to foliar-applied solutes may result from osmotic shock. Brief osmotic shocks adversely affect the integrity of the plasma membrane in both bacterial and plant cells.[78-86] Plant responses to osmotic shock include:

1. Decreased dry weight and inhibited growth[84]
2. Inhibition of auxin-induced growth responses[80]
3. Alterations in transmembrane electrode potentials[79,81]
4. Accelerated leakage of cell contents from disks of beetroot segments,[83] coleoptiles,[79] and corn leaf sections[78]

Neumann[78] concluded that membrane leakage induced by osmotic shock could be a cause of burn damage when high concentrations of osmotically active solutes are applied to the leaves. Membrane-damaging concentrations ranged from 3.5 to 31 g/ℓ, depending on fertilizer assayed. Differences in rates of both transcuticular and transmembrane equilibration were presumed to be involved in the differences in damage response to different solutes.

3. Toxic Damage
In a subsequent study, the threshold concentrations at which various fertilizer sources of nitrogen, phosphorus, potassium, and sulfur caused burn damage, when applied under standard conditions to intact corn leaf surfaces, was determined. The results (Table 13) did not support the concept of a single plasmolytic mechanism of damage. For example, whereas droplets of KH_2PO_4 induced damage at 0.50 M (68 g/ℓ), as might be expected for osmotic damage, K_2HPO_4 became damaging at 0.05 M or 9 g/ℓ. Moreover, penetration of the leaf cuticle and translocation away from the site of application were equivalent for ^{32}P-labeled samples of either salt. These results suggested that all osmotically active solutes can induce plasmolytic damage if they penetrate into intact leaves at sufficiently high concentrations, whereas some, with aqueous pHs below 4.0 or above 8 (e.g., K_2HPO_4), can induce toxic damage at far lower concentrations. Neumann and Golab[88] went on to show that exposure of leaf cells to hypotonic K_2HPO_4 solutions (but not KH_2PO_4), at 0.05 and 0.10 M for 90 min, reduced the capacity of these cells to withstand subsequent exposure to osmotic stress. Stress-induced damage (i.e., leakage of cell constituents) in K_2HPO_4 pretreated leaf cells, was 87% higher than in water and KH_2PO_4 pretreated controls. Interestingly, the necrotic

Table 13

**CONCENTRATIONS AT WHICH 20-$\mu\ell$ DROPLETS
OF FERTILIZER SOLUTIONS APPLIED TO
INTACT SURFACE OF MAIZE LEAVES BEGIN TO
INDUCE DAMAGE**

Fertilizer applied	Damaging concentration (molarity)	Solution pH
KH_2PO_4	0.50	4.4
K_2HPO_4	0.05	9.3
K Poly P	0.04[a]	8.7
NH_4 Poly P (powdered)	0.50[a]	5.0
NH_4 Poly P (liquid)	0.10[a]	6.9
$NH_4H_2PO_4$	0.50	4.4
$(NH_4)_2HPO_4$	0.10	8.3
$(NH_4)_3PO_4$	0.10	8.6
Urea phosphate	0.03	2.2
Urea	0.40	6.0
NH_4NO_3	0.40	4.5
KNO_3	0.40	5.3
K_2SO_4	0.50	6.1
$(NH_4)_2SO_4$	0.40	5.5

[a]   Calculated as moles orthophosphate liter^{-1}; powdered source 90% of phosphorus as polyphosphates; liquid source 50% of phosphorus as polyphosphates.

Adapted from Neumann, P. M., Ehrenreich, Y., and Golab, Z., *Agron. J.*, 73, 979, 1981.

damage induced by applications of 0.05 M K_2HPO_4 to intact leaves could be inhibited by maintaining the treated plants under high humidity (low stress) conditions. The manner in which K_2HPO_4 selectively weakens the plasma membrane is not clear. Possibly, it is related to the intracellular buffering capacity of the cells[89] being overloaded by excessive uptake of alkaline KH_2PO_4.

Foliar sprays of salt solutions can also cause other, nonnecrotic, damage. Thus, Swietlik et al.[90] found that foliar sprays of a complete nutrient solution led to reduced photosynthesis and stomatal conductance in apple seedlings. They subsequently showed that the inhibitory effect was caused by $CaCl_2$.[91] These authors speculated that an excess of anion (Cl^-) over cation (Ca^{2+}) uptake led to leaf cell acidification and that lowered stomatal guard cell pH led to stomatal closure.

These data reveal the complexities involved in attempting to evaluate possible damaging effects of foliar-applied solutes. Calcium is believed to have a direct role in promoting photosynthesis[92] and calcium deficiencies inhibit photosynthesis.[93] Nevertheless, foliar applications of $CaCl_2$ at 0.025 M inhibited apple leaf photosynthesis by inducing stomatal closure and via direct adverse effects on mesophyll cell activity.

One common thread linking the damaging effects of various salts applied at low (nonplasmolytic) concentrations is the hypothesis that the buffer capacity of the cell pH stat[89] may be overridden by excessive and unbalanced solute influx into leaf cells.

V. CONCLUDING REMARKS

Much evidence supports the contention that declines in the flux of essential solutes in the cells of mature leaves may be responsible for ontogenetic declines in photosynthetic capacity.

This chapter presents several lines of experimental evidence suggesting that age and/or deficiency-related declines in the photosynthetic capacity of individual leaves can be reversed or reduced by leaf uptake of exogenous chemicals such as mineral nutrients, cytokinins, and CO_2. However, it is only in the case of soil-induced micronutrient deficiencies that economical control of leaf development via foliar sprays is achieved in current agricultural practice. The results reviewed here suggest that agricultural benefits will also derive from ongoing research into CO_2 fertilization and the use of senescence-retarding foliar sprays.

REFERENCES

1. **Turgeon, R.,** Termination of nutrient import and development of vein loading capacity in albino tobacco leaves, *Plant Physiol.,* 76, 45, 1986.
2. **Thorpe, M. R. and Lang, A.,** Control of import and export of photosynthate in leaves, *J. Exp. Bot.,* 34, 231, 1983.
3. **Besford, R. T., Withers, A. C., and Ludwig, L. J.,** Ribulose bisphosphate carboxylase activity and photosynthesis during leaf development in the tomato, *J. Exp. Bot.,* 36, 1530, 1985.
4. **Makino, A., Mae, T., and Ohiro, K.,** Relations between nitrogen and ribulose-1.5-bisphosphate carboxylase in rice leaves from emergence through senescence, *Plant Cell Physiol.,* 25, 429, 1984.
5. **Evans, J. R.,** Nitrogen and photosynthesis in the flag leaf of wheat (*Triticum aestivum* L.), *Plant Physiol.,* 72, 297, 1983.
6. **Pate, J. S. and Atkins, C. A.,** Xylem and phloem transport and the functional economy of carbon and nitrogen of a legume leaf, *Plant Physiol.,* 71, 835, 1983.
7. **Dale, J. E.,** The growth of leaves, *Inst. of Biology Publication 137,* Edward Arnold, London, 1982, 50.
8. **Jenkins, G. I. and Woolhouse, H. W.,** Photosynthetic electron transport during senescence of the primary leaves of *Phaseolus vulgaris* L. I. Non cyclic electron transport, *J. Exp. Bot.,* 32, 467, 1981.
9. **Friedrich, J. W. and Huffaker, R. C.,** Photosynthesis, leaf resistances. Ribulose-1,5-bisphosphate carboxylase degradation in senescing barley leaves, *Plant Physiol.,* 65, 1103, 1980.
10. **Davis, S. D. and McCree, K. J.,** Photosynthetic rate and diffusion conductance as a function of age in leaves of bean plants, *Crop. Sci.,* 18, 280, 1978.
11. **Catsky, J., Ticha, I., and Solerova, J.,** Ontogenetic changes in the internal limitations to bean-leaf photosynthesis. I. Carbon dioxide exchange and conductances for carbon dioxide transfer, *Photosynthetica,* 10, 394, 1976.
12. **Thomas, H. and Stoddart, J. L.,** Leaf senescence, *Annu. Rev. Plant Physiol.,* 31, 83, 1980.
13. **Noodén, L. D.,** Senescence in the whole plant, in *Senescence in Plants,* Thimann, K. V., Ed., CRC Press, Boca Raton, Fla., 1980, chap. 10.
14. **Bunce, J. A.,** Measurements and modelling of photosynthesis in field crops, *CRC Crit. Rev. Plant Sci.,* 4, 47, 1986.
15. **Gifford, R. M., Thorne, J. H., Hitz, W. O., and Giaquinta, R. T.,** Crop productivity and photoassimilate partitioning, *Science,* 225, 801, 1984.
16. **Marney, J. R., Frye, K. E., and Guinn, G.,** Relationship of photosynthesis rate to growth and fruiting of cotton, soybean, sorghum and sunflower, *Crop Sci.,* 18, 259, 1978.
17. **Neumann, P. M. and Stein, Z.,** Relative rates of delivery of xylem solute to shoot tissues: possible relationship to sequential senescence, *Physiol. Plant.,* 62, 390, 1984.
18. **Davis, S. D., Van Bavel, C. H. M., and McCree, K. J.,** Effect of leaf aging upon stomatal resistance in bean plants, *Crop Sci.,* 17, 640, 1977.
19. **Wittenbach, V. A.,** Effect of pod removal on leaf photosynthesis and soluble protein composition of field-grown soybeans, *Plant Physiol.,* 73, 121, 1983.
20. **Peoples, M. B., Beilharz, V. C., Waters, S. P., Simpson, R. J., and Dalling, M. J.,** Nitrogen redistribution during grain growth in wheat. II. Chloroplast senescence and the degradation of ribulose-1:5-bisphosphate carboxylase, *Planta,* 149, 241, 1980.
21. **Neumann, P. M. and Stein, Z.,** Xylem transport and the regulation of leaf metabolism, *What's New Plant Physiol.,* 14, 33, 1983.
22. **Neumann, P. M. and Noodén, L. D.,** Pathway and regulation of phosphate translocation to the pods of soybean explants, *Physiol. Plant.,* 60, 166, 1984.
23. **Layzell, D. B., Pate, J. S., Atkins, C. A., and Canvin, D. T.,** Partitioning of carbon and nitrogen and the nutrition of root and shoot apex in a nodulating legume, *Plant Physiol.,* 67, 30, 1981.

24. **Ozuna, R., Yera, R., Ortega, K., and Tallmann, G.,** Analysis of guard cell viability and action in senescing leaves of *Nicotinia glauca* (Graham), tree tobacco, *Plant Physiol.,* 79, 7, 1985.
25. **Harris, G. C., Cheesbrough, J. K., and Walker, D. A.,** Measurement of CO_2 and H_2O vapour exchange in spinach leaf discs. Effects of orthophosphate, *Plant Physiol.,* 71, 102, 1983.
26. **Shaner, D. L. and Boyer, J. S.,** Nitrate reductance activity in maize *(Zea mays* L.) leaves. I. Regulation by nitrate flux, *Plant Physiol.,* 58, 449, 1976.
27. **Neumann, P. M. and Noodén, L. D.,** Characterization of leaf senescence and pod development in soybean explants, *Plant Physiol.,* 72, 182, 1983.
28. **Martin, C. and Thimann, K. V.,** Role of protein synthesis in the senescence of leaves. I. The formation of protease, *Plant Physiol.,* 49, 64, 1972.
29. **Tanada, T.,** Role of boron in the far-red delay of nyctinastic closure of *Albizzia pinnules, Plant Physiol.,* 70, 320, 1982.
30. **Makino, A., Mae, T., and Ohiro, K.,** Effect of nitrogen phosphorus or potassium on the photosynthetic rate and ribulose 1,5-bis-phosphate carboxylase content in rice leaves during expansion, *Soil Sci. Plant Nutr.,* 30, 63, 1984.
31. **Longstreth, D. J. and Nobel, P. S.,** Nutrient influences on leaf photosynthesis. Effects of nitrogen phosphorus and potassium for *Gossypium hirsutum* L. *Plant Physiol.,* 65, 541, 1980.
32. **Oxman, A. M., Goodman, P. J., and Cooper, J. P.,** The effects of nitrogen phosphorus and potassium on rates of growth and photosynthesis of wheat, *Photosynthetica,* 11, 66, 1977.
33. **Natr, L.,** Influence of mineral nutrients on photosynthesis of higher plants, *Photosynthetica,* 6, 80, 1972.
34. **Epstein, E.,** *Mineral Nutrition of Plants: Principles and Perspectives,* John Wiley & Sons, New York, 1972.
35. **Neumann, P. M. and Chamel, A.,** Comparative phloem mobility of nickel in non-senescent plants, *Plant Physiol.,* 81, 689, 1986.
36. **Luettge, U. and Higginbotham, N.,** *Transport in Plants,* Springer-Verlag, New York, 1979, 72.
37. **Crafts-Brandner, S. J., Below, F. E., Wittenbach, V. A., Harper, J. E., and Hageman, R. H. D.,** Differential senescence of maize hybrids following ear removal. II. Selected leaf, *Plant Physiol.,* 74, 368, 1984.
38. **Vann, D. R., Fletcher, J. S., Acchireddy, N. R., and Keevers, L.,** Influence of partial defoliation of green pepper on the senescence, growth and nitrate reductase of the remaining leaf, *Plant Soil,* 91, 357, 1986.
39. **Neumann, P. M. and Stein, Z.,** Ion supply capacity of roots in relation to rejuvenation of primary leaves in vivo, *Physiol. Plant.,* 67, 97, 1986.
40. **Ness, P. J. and Woolhouse, H. W.,** RNA synthesis in phaseolus chloroplasts. II. Ribonucleic acid synthesis in chloroplasts from developing and senescing leaves, *J. Exp. Bot.,* 31, 235, 1980.
41. **Carmi, A. and Koller, D.,** Regulations of photosynthetic activity in the primary leaves of bean *(Phaseolus vulgaris* L.) by materials moving in the water conducting system, *Plant Physiol.,* 64, 285, 1979.
42. **Hodgkinson, K. C.,** Influence of partial defoliation on photosynthesis, photorespiration and transpiration by lucerne leaves of different ages, *Aust. J. Plant Physiol.,* 1, 561, 1974.
43. **Gifford, R. M. and Marshall, C.,** Photosynthesis and assimilate distribution in *Lolium multiflorum Lam* following differential tiller defoliation, *Aust. J. Biol. Sci.,* 26, 517, 1973.
44. **Wareing, P. F., Khalifa, M. M., and Treharne, K. J.,** Rate-limiting processes in photosynthesis at saturating light intensities, *Nature,* 220, 453, 1968.
45. **Itai, C. and Vaadia, Y.,** Cytokinin activity in water stressed shoots, *Plant Physiol.,* 47, 87, 1971.
46. **Hsiao, T. and Bradford, K. J.,** Physiological consequences of cellular water deficits, in *Limitations to Efficient Water use in Crop Production,* Taylor, H. M., Jordon, W. R., and Sinclair, T. R., Eds., American Society of Agronomy, Madison, Wis., 1983, chap. 6A.
47. **Van Staden, J. and Carmi, A.,** The effects of decapitation on the distribution of cytokinins and growth of *Phaseolus vulgaris* plants, *Physiol. Plant.,* 55, 39, 1982.
48. **Colbert, K. A. and Beever, J. E.,** Effect of disbudding on root cytokinin export and leaf senescence in tomato and tobacco, *J. Exp. Bot.,* 32, 121, 1981.
49. **Palmer, M. V., Horgan, R., and Wareing, P. F.,** Cytokinin metabolism in *Phaseolus vulgaris* L. I. Variations in cytokinin levels in leaves of decapitated plants in relation to lateral bud outgrowth, *J. Exp. Bot.,* 32, 1231, 1981.
50. **Neumann, P. M. and Noodén, L. D.,** Interaction of mineral and cytokinin supply in control of leaf senescence and seed growth in soybean explants, *J. Plant Nutr.,* 6, 735, 1983.
51. **Naito, K., Nsegumo, S., Furuya, K., and Suzuki, H.,** Effect of benzylademe on RNA and protein synthesis in intact bean leaves at various stages of ageing, *Physiol. Plant.,* 52, 343, 1980.
52. **Noodén, L. D., Kahanak, G. M., and Okatan, Y.,** Prevention of monocarpic senescence in soybeans with auxin and cytokinin: an antidote for self destruction, *Science,* 206, 841, 1979.
53. **Wittenbach, V. A.,** Induced senescence of intact wheat seedlings and its reversibility, *Plant Physiol.,* 59, 1039, 1977.

54. **Thimann, K. V., Tetley, R. M., and Thanh, T. V.,** The metabolism of oat leaves during senescence. II. Senescence in leaves attached to the plant, *Plant Physiol.,* 54, 859, 1974.

55. **Fletcher, R. A.,** Retardation of leaf senescence by benzyladenine in intact bean plants, *Planta,* 89, 1, 1969.

56. **Adedipe, N. O., Hunt, L. A., and Fletcher, R. A.,** Effects of benzyladenine on photosynthesis, growth and senescence of the bean plant, *Physiol. Plant.,* 25, 151, 1971.

57. **Garcia, L. and Hanway, J. J.,** Foliar fertilization of soybeans during the seed filling period, *Agron. J.,* 68, 653, 1976.

58. **Sinclair, T. R. and Wit, C. T.,** Photosynthetate and nitrogen requirements for seed production by various crops, *Science,* 189, 565, 1975.

59. **Neumann, P. M.,** Late season foliar fertilization with macronutrients — is there a theoretical basis for increased seed yields?, *J. Plant Nutr.,* 5, 1209, 1982.

60. **Phillips, D. A., Pierce, R. O., Edie, S. A., Foster, K. W., and Knowles, P. F.,** Delayed leaf senescence in soybean, *Crop Sci.,* 24, 518, 1984.

61. **Abu-Shakru, S. S., Phillips, D. A., and Huffaker, R. C.,** Nitrogen fixation and delayed leaf senescence in soybeans, *Science,* 199, 973, 1978.

62. **Swietlik, D. and Faust, M.,** Foliar nutrition of fruit crops, *Hortic. Rev.,* 6, 278, 1984.

63. **Kannan, S.,** Foliar absorption and transport of inorganic nutrients, *CRC Crit. Rev. Plant Sci.,* 4, 341, 1986.

64. **Neumann, P. M. and Prinz, R.,** Foliar iron spray potentiates growth of seedlings on iron free media, *Plant Physiol.,* 55, 988, 1975.

65. **Eskew, D. H., Welch, R. M., and Cary, E. E.,** Nickel: an essential micronutrient for legumes and possibly all higher plants, *Science,* 222, 621, 1983.

66. **Klucas, R. V., Hanus, F. J., Russel, S. A., and Evans, H. J.,** Nickel: a macronutrient element for hydrogen-dependent growth of *Rhizobium japonicum* and for expression of urease activity in soybean leaves, *Proc. Natl. Acad. Sci. U.S.A.,* 80, 2253, 1983.

67. **Welch, R. M.,** The biological significance of nickel, *J. Plant Nutr.,* 3, 345, 1981.

68. **Havelka, U. D., Wittenbach, V. A., and Boyle, M. G.,** CO_2-enrichment effects on wheat yield and physiology, *Crop Sci.,* 24, 1163, 1984.

69. **Manning, D. T., Chen, T. M., Campbell, A. J., Tolbert, N. E., and Smith, E. W.,** Effects of chemical treatments upon photosynthetic parameters in soybean seedlings, *Plant Physiol.,* 76, 1055, 1984.

70. **Havelka, U. D., Ackerson, R. C., Boyle, M. G., and Wittenbach, V. A.,** CO_2-enrichment effects on soybean physiology. I. Effects of long term CO_2 exposure, *Crop Sci.,* 24, 1146, 1984.

71. **Ackerson, R. C., Havelka, U. D., and Boyle, M. G.,** CO_2 enrichment effects on soybean physiology. II. Effects of stage specific CO_2 exposure, *Crop Sci.,* 24, 1150, 1984.

72. **Hardman, L. L. and Brun, W. A.,** Effect of atmospheric carbon dioxide enrichment at different developmental stage on growth and yield components of soybeans, *Crop Sci.,* 11, 886, 1971.

73. **Rogers, H. H., Sionit, N., Cure, J. D., Smith, J. M., and Bingham, G. E.,** Influence of elevated carbon dioxide on water relations of soybeans, *Plant Physiol.,* 74, 233, 1984.

74. **Gris, E.,** Nouvelles experiences sur l'action des composes ferrigineux solubles appliques á la vegetation et specialement de la chlorose et la debilité des plantes, *Compte Rendu (Paris),* 19, 118, 1844.

75. **Neumann, P. M. and Prinz, R.,** Evaluation of surfactants for use in the spray treatment of iron chlorosis in citrus trees, *J. Sci. Food Agric.,* 25, 221, 1974.

76. **Wallihan, E. F., Embleton, T. W., and Sharpless, R. G.,** Response of chlorotic citrus leaves to iron sprays in relation to surfactants and stomatal apertures, *Proc. Am. Soc. Hortic. Sci.,* 85, 210, 1964.

77. **Reed, D. W. and Tukey, H. B., Jr.,** Effect of pH on foliar absorption of phosphorus compounds by chrysanthemum, *J. Am. Soc. Hortic. Sci.,* 103, 337, 1978.

78. **Neumann, P. M.,** Rapid evaluations of foliar fertilizer induced damage: NPKS on corn, *Agron. J.,* 71, 598, 1979.

79. **Rubinstein, B., Mahar, P., and Tattar, T.,** Effects of osmotic shock on some membrane regulated events of oat coleoptile cells, *Plant Physiol.,* 59, 365, 1977.

80. **Rubinstein, B.,** Osmotic shock inhibits auxin-stimulated acidification and growth, *Plant Physiol.,* 53, 369, 1977.

81. **Racusen, R. H., Kinnersley, A. M., and Galston, A. W.,** Osmotically induced changes in electrical properties of plant protoplast membranes, *Science,* 198, 405, 1977.

82. **Amar, L. and Reinhold, L.,** Loss of membrane transport ability in leaf cells and release of protein as a result of osmotic shock, *Plant Physiol.,* 51, 620, 1973.

83. **Greenway, H.,** Effects of slowly permeating osmotica on metabolism of vacuolated and nonvacuolated tissues, *Plant Physiol.,* 46, 254, 1970.

84. **Greenway, H. and Leahy, M.,** Effects of rapidly and slowly permeating osmotica on metabolism, *Plant Physiol.,* 46, 259, 1970.

85. **Heppel, L. A.,** Selective release of enzymes from bacteria, *Science,* 156, 1451, 1967.
86. **Heppel, L. A.,** The effect of osmotic shock on release of bacteria proteins and on active transport, *J. Gen. Physiol.,* 54, 955, 1969.
87. **Neumann, P. M., Ehrenreich, Y., and Golab, Z.,** Foliar fertilizer damage to corn leaves: relation to cuticular penetration, *Agron. J.,* 73, 979, 1981.
88. **Neumann, P. M. and Golab, Z.,** Comparative effects of mono- and dipotassium phosphates on cell leakiness in corn leaves, *J. Plant Nutr.,* 6, 275, 1983.
89. **Smith, F. A. and Raven, J. A.,** Intracellular pH and its regulation, *Annu. Rev. Plant Physiol.,* 30, 289, 1979.
90. **Swietlik, D., Faust, M., and Korcak, R. F.,** Effect of mineral nutrient sprays on photosynthesis and stomatal opening of water stressed and untreated apple seedlings. I. Complete nutrient sprays, *J. Am. Soc. Hortic. Sci.,* 107, 563, 1982.
91. **Swietlik, D., Bunce, J. A., and Miller, S. S.,** Effect of foliar application of mineral nutrients on stomatal aperture and photosynthesis in apple seedlings, *J. Am. Soc. Hortic. Sci.,* 109, 306, 1984.
92. **Brand, J. and Becker, D. W.,** Evidence for direct roles of calcium in photosynthesis, *J. Bioenerg. Biomemb.,* 16, 239, 1984.
93. **Veierskov, B. and Meravy, L.,** Changes in carbohydrate composition in wheat and pea seedlings induced by calcium deficiency, *Plant Physiol.,* 79, 315, 1985.

Chapter 4

VARIABILITY IN RESPONSES OF DEVELOPING PLANT TISSUES TO APPLIED CHEMICALS — THE CONCEPT OF TARGET CELLS

Daphne J. Osborne

TABLE OF CONTENTS

I. INTRODUCTION

A. The Role of Man

Man has always manipulated the living things that surround him, either to increase their value to him or to reduce their interference with his life style. Early man, by virtue of his limited numbers and primitive technology, made small inroads only into the structure of the environment, but as he progressed from hunter-gatherer to form stable sedentary societies with organized farming, his impact grew prodigiously. Nonetheless, until medieval times little effort was made to modify the way plants grew; the questions of where the crop was grown and how the land was managed were paramount. In Great Britain we can still see this legacy in the ridge and furrow profiles of many pastures that were once part of open-field strip farming. Many of the hedgerows planted early in the 16th century following the inclosure of common land for private use are still in existence today. As long as relatively small fields of similar species were cultivated in any one region, the grower had to be content with the yield the land provided — the best he could do was to irrigate by simple (or sometimes complex) means and to add animal waste as a fertilizer.

B. The Eradication of Pathogens

As areas of cultivation of single crops enlarged, the build-up of pathogens increased in importance and man quickly learned to refrain from farming the same annual crop in the same ground for several years in succession. However, with the development of large colonial plantations of long-lived crops, the lack of effective means for protecting the plants against pathological infections or insect predation became pressing in the extreme. The cessation of coffee growing in Ceylon at the end of the 19th century resulted from the inability to control the massive spread of the rust, *Hemileia vastatrix*. The elimination of rubber from large areas of Africa and South America and cloves from Zanzibar are other examples. The destruction of the staple potato crop in Ireland by the fungal pathogen *Phytophthora infestans*, and the famine that ensued, occurred less than 150 years ago. Then the only simple recourse was to a general application of flowers of sulfur, copper, or arsenic compounds to the foliage while the hope for the future centered on the efforts of plant breeders to produce pathogen-resistant varieties. Now genetic engineers have joined the plant breeders, and an armory of complex organic chemicals are available for plant protection.

C. The Eradication of Weeds

The grower has always recognized the importance of eradicating weeds that compete with his crop. As the cost of hand labor increased into the 20th century, the use of chemicals as weedkillers became progressively more acceptable. The method of application was similar to that for pathogen control. The chemical was watered or sprayed onto the foliage with the aim of either a quick total kill to clear all vegetation or a quick selective kill that would leave the crop plants essentially unscathed. The basis of selectivity was more dependent upon the relative amount of the chemical that was retained on the leaf surface than on any inherent selectivity of the chemical for disrupting biochemical pathways in the cells of the recipient plants. In this way, for example, the French pioneered the use of sulfuric acid as a selective weedkiller of Charlock in cereal crops and by 1937 this procedure was common in many European countries. By then, organic compounds has also gained a place in agriculture and despite the intense yellow color imparted to everything the chemical reached, salts of dinitro-orthocresols were sprayed onto fields of cereals to preferentially arrest the growth of broad-leaved weeds. Even with these unsophisticated chemicals which could kill the aerial parts in a matter of hours, it was evident that destruction of the weed and successful selective survival of the crop was subject to many modifying environmental factors, particularly those of temperature and light. But, most especially, it was clear that the stage of growth and development of both the weed and crop were critical.

D. Chemicals with Specificity

The start of the great revolution in the manipulation of plants by chemical means came in the 1940s as a direct outcome of basic research devoted to elucidating the mechanism by which auxin (indole-3-acetic acid [IAA]), a naturally produced plant growth regulator, controlled the extension growth of cells and hence modified the form and development of roots and shoots. Many synthetic analogs of IAA were either equally or more effective in causing these modifications. When it was shown in 1940 by Slade, Templeton, and Sexton that foliar sprays of some of these growth-promoting substances would selectively kill annual broad-leaved weeds but similarly treated cereals remained unharmed, the era of biologically targeted plant manipulation had arrived.[1]

II. TARGETED RESPONSES

A. Chemicals and the Plant

When any chemical is applied to a plant, whether it be a toxic killer like sulfuric acid, a fungicide such as the copper-based Bordeaux mixture, a nutrient, or a fertilizer, the morphological factors that determine retention and penetration into the plant and physiological factors that determine translocation and metabolism within the tissues all interact to influence the result of the treatment in quantitative terms. These important aspects of foliar treatments are dealt with in detail in other chapters of this book. In contrast, growth-regulating chemicals provide an additional and qualitative dimension to plant manipulation and the expression of the genome. The life span of a leaf, the source-sink movement of assimilates, the shedding of plant parts, the expression of sex, parthenocarpy, and the production of cell-specific protein determinants may all be controlled by chemicals with the requisite regulatory properties. Between the 1940s and the 1980s growth-regulatory chemicals have become progressively more tailor-made to fulfill specific requirements in specific plants and the products used have reached high standards of reproducibility in quality and activity. The performance of the plant in response to an applied chemical is, however, considerably less predictable. In particular, the responses to hormonal analogs may show a marked variation from one group of plants to another, from one plant part to another, and from one plant cell to another.[2]

Why should this be?

B. The Target-Cell Concept

The concept that there is a specific but progressively changing target status for each cell throughout the life-span of its development is central to present thinking on the differentiation and organization of plant form and function.[3] Every living cell in a plant is seen as possessing a specific target status with respect to the numerous environmental, hormonal, and chemical signals to which it is exposed. This means that if a cell has the competence to recognize a signal and respond to that signal in a specific way, then it is a target cell for that signal. If it is competent to recognize the signal but fails to respond, it is a repressed target cell. But if the cell lacks the competence to recognize the signal, then response to that signal is precluded and the cell cannot then be a target cell for that specific recognition and response.

Each cell reflects its state of competence in precise molecular terms detectable in the complement of proteins or immunologically antigenic determinants that express the operating condition of the genome.[4]

In the complex structure of the whole plant, cells with different gene expression and, hence, different molecular expression of proteins, nucleic acids, or carbohydrate moities are differentiated in a precisely restricted and organized way. In turn, each cell responds to the variety of external stimuli or the concert of internal hormonal signals in a manner determined by its target status or competence for specific signal recognition and response.

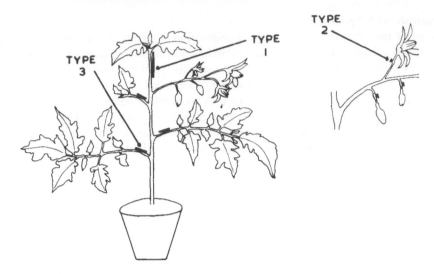

FIGURE 1.   Positional differentiation of three different types of ethylene responsive target cells in the tomato plant. Immature cells of the extending shoot (Type 1); cells of the abscission zones at base of the flower pedicels (Type 2); epinasty-associated cells on the adaxial flanks at the base of the leaf petioles (Type 3).

C. Ethylene Target Cells

An example of the different cell-specific responses that can result from a similar hormonal stimulus can be readily demonstrated when a tomato plant is exposed to ethylene gas (Figure 1). The responses that ensue are evidence of the broad differences in target-cell types that can be encountered in the aerial shoot.[5]

1. Type 1 Cells

The cells of the extending region below the apical meristem will exhibit either a reduced rate of extension growth or cease extension altogether when the plant is exposed to ethylene. These immature cells will, however, continue to expand in volume — but more slowly — in the lateral direction, resulting in a swollen and stunted shoot apex. This growth response in the presence of ethylene is typical of the subapical regions of most higher plant roots and shoots and is associated with an ethylene-induced reorientation of the direction of deposition of the cellulosic microfibrils that provide the strength to the cell wall. Deposition of the fibrils is normally in the horizontal plane with respect to the vertical shoot, so that the girdling of the wall preferentially permits an anisotropic extension of the cell in the longitudinal direction. Under the influence of ethylene, however, new microfibrils are deposited preferentially along the vertical axis of the cell wall, hence the laterally permissive anisotropy of expansion that subsequently follows. A change in the direction of cell expansion is one of the most commonly encountered results of an ethylene experience and at similar concentrations of ethylene, shoots of dicotyledenous plants show a considerably greater reaction to the signal than most monocotyledons. The reason for this is currently unclear. Cells that correspond to this target status all exhibit an auxin enhancement of elongation as exemplified by the classical experiments of auxin-induced cell elongation found in segments of pea shoots or in the coleoptiles of cereal seedlings.[6] Cells that react in this way to ethylene and to auxin have been collectively called Type 1.[5]

2. Type 2 Cells

The response of those cells that comprise the abscission zone at the base of the flower pedicel (Figure 1) is quite different from that shown by the immature Type 1 cells of the shoot apex. In the presence of ethylene, target cells of an abscission zone can increase their overall size and at least one new gene expression is elicited in the form of a zone cell-specific isozyme of β-1:4-glucanhydrolase. This and other enzymes are produced and secreted across the plasma membrane and into the cell wall and are responsible for the degradation of the polysaccharides of the middle lamellae between the neighbor cells of the zone. In this way, the separation of the cells, one from another, and the shedding of the flower occur only at the very precise location delimited by the positionally differentiated abscission zone cells and nowhere else. Ethylene target cells, collectively classified as Type 2, exhibit a further and remarkable diagnostic characteristic in that they do not enlarge in response to auxin. In fact, auxins inhibit the ethylene-induced cell enlargement, repress the expression of the ethylene-inducible β-1:4-glucanhydrolase, and block the secretion of this and other enzyme proteins to the wall.[7] Investigations in other species such as the bean, *Phaseolus vulgaris* and the elderberry, *Sambucus nigra* have shown that the protein complement of the abscission zone target cells contains a small number of proteins that are specific to the zone cells and absent from nonzone cells. The presence of these particular proteins can therefore be used as a biochemical "marker" of the target status of the Type 2 cell and represents a selective and specific gene expression for the cell type.[7]

3. Type 3 Cells

We can recognize a third type of ethylene-responsive target cell in the leaf petioles of a number of land species including the tomato plant (Figure 1). In these, a drooping of the petiole (epinasty) occurs when the plants are exposed to either ethylene, auxins, or the conditions that lead to abnormally high production of ethylene in the shoot, e.g., waterlogging. This downward bending of the petiole results from the elongation of a specific group of cells located on the flanks of the upper side of the base of the petiole. These cells will elongate normally under the control of auxin, in parallel with the other cells of the petiole, but in the presence of ethylene they undergo a further and additional growth response which leads to the downward curvature of the petiole base. These specialized and unusual cells, which enlarge and elongate in response to both auxin and to ethylene have been classed as Type 3.[5]

Relatively few cells with these characteristics are differentiated in land plants, but they make up the major proportion of the cells in the petioles, internodes, and flower stalks of most semiaquatic or amphibious plants such as the watercress, *Rorippa nasturtium-aquaticum*, the frogbit, *Hydrocharis morsus-ranae*, and seedling paddy rice, *Oryza sativa*. Fronds of the water fern, *Regnellidium diphyllum*, a plant that covers the muddy banks of vast stretches of the Amazon River, and the petioles of water lilies such as *Nymphoides peltata* exhibit impressive examples of Type 3 cell growth.[8] When the vegetative parts become submerged, the ethylene normally produced by the plants starts to accumulate within the tissues because of the lower rate of diffusion of the gas into water as compared to air. The concentrations reached in the internal air spaces become high enough to initiate the typical ethylene-induced cell elongation. When the leaves again reach the water surface the accumulated ethylene is dispersed into the air and the growth rate declines to normal, but the greatly elongated rachis or petiole remains as evidence of the earlier Type 3 cell response.[8]

These three ethylene-regulated target-cell types may be considered extreme examples of the responses that can result when plant foliage is exposed to a particular growth regulator, but they serve to illustrate the point that not every cell will or can react in the same way to an applied chemical. Nor can all cells respond in the same way to the stimuli that they encounter daily from the external environment.

D. Diversity of Target-Cell Types

Among different species of plants, cells that show apparently similar target characteristics, e.g., abscission cells, do not necessarily achieve competence for the same signal recognition and response.

1. Abscission Zone Cells of Dicotyledonous Plants

Abscission zone cells of dicotyledons all exhibit the characteristics of Type 2 cells in that they can enlarge and separate in response to ethylene, and the ethylene response is repressed by auxins. However, abscission zone cells of vegetative and flowering parts can exhibit certain differences. This is of considerable importance for the success of the aerial sprays that enhance cell separation and facilitate fruit harvesting by mechanical means and is particularly important in collection procedures from evergreen trees such as olive or citrus. The goal of selectively loosening the fruit, but not loosening the leaves, in these species is clearly critical to efficient harvesting and the overall welfare of the tree.

How the differentiation program of abscission zone cells may differ in reproductive tissue as distinct from leaf tissue has yet to be determined, but analysis of the abscission performance of zones from vegetative and reproductive parts suggests that the responses to abscission-regulating signals could be sufficiently different to permit their selective activation. Thus, experiments with *Sinapis alba* have identified the appearance of two new proteins in the shoot meristem and the disappearance of a third when vegetative plants receive the photoinductive light-dark regime that leads to flowering.[9] This change in protein complement becomes irreversibly entrained for the reproductive condition within 20 hr of photoinduction and can be considered as "marking" the flowering competent meristem. Abscission zones differentiated in the leaves and floral structures of the photoinduced plants are therefore distinct in this respect from those of the leaves from noninduced plants and should, in theory, exhibit certain differences in response to abscission signals.

Whereas all abscission zone target cells of dicotyledons respond to ethylene, there is already good evidence that leaf and fruit zones do not react to the signal with equal speed and may also separate in slightly different ways.[10] In horticulture therefore, there is considerable room for the scientist to expand his understanding of such differences in gene expression and provide new and subtle means for causing the removal of the required specific plant parts in preference to others.

2. Abscission Zone Cells of the Graminae

Whereas abscission zone cells in all the dicotyledonous plants so far studied can be induced to separate by the appropriate concentrations of ethylene, those of the fruit pedicels of the Graminae are not controlled by the same signal; ethylene does not initiate the separation process.[11] The fruits of wild grasses are shed from the parent plant by the loss of cell-to-cell adhesion at a precisely delineated position at the fruit base. The cells that separate are differentiated soon after anthesis as a plate between the fruit base and the fruit pedicel, but the signal for separation is transmitted from the maturing fruit only when the caryopsis begins to lose water and the glumes start to senesce and desiccate. Interestingly, the progressive dehydration leads to a reduction in the level of ethylene production, but it is also the condition in which the level of abscisic acid (ABA) in the fruit increases. In vitro experiments with explants of *Avena fatua* fruits from which the caryopsis and glumes were removed, leaving the fruit base, the abscission zone, and the pedicel, have confirmed that ethylene is not the signal to induce abscission; they have also demonstrated that another hormone, ABA, is highly effective in this respect.[11] On this evidence it appears that the abscission zone cells of the Graminae are target cells for ABA but not for ethylene and that ABA acts as the derepressor of the genes responsible for the activity of enzymes that degrade the middle-lamella polysaccharides at the zone.

The eventual loss of cell-to-cell adhesion of the fruit-pedicel junction is thus a similar type of event to that in dicotyledons, but is controlled by a different signal recognition and response. In many respects it is not unexpected that ethylene should not be the initiator of shedding in the Graminae as cells that become dehydrated also show impaired membrane organization and readily leak solutes across the lipid bilayer of the plasma membrane. Ethylene biosynthesis is dependent upon a precise integration of the relevant enzymes in a functional membrane and production has been shown to fall when the membrane integrity is impaired by temperature extremes or dehydration. In contrast, abscisic acid levels rise when the water potential of a cell falls and for seeds that must be shed essentially dry it can be considered an evolutionary advantage to have the abscission zone comprised of target cells for abscisic acid rather than for ethylene.[12]

3. Role of Abscisic Acid

Although there is so much evidence to show the effectiveness of ethylene as an initiator and accelerator of abscission in dicotyledons, it should be pointed out here that abscisic acid is also a potent accelerator of separation in most excised abscission zones. When the much favored "explant" segment which contains an abscission zone and tissue both distal and proximal to the zone is given ABA in a microdrop applied to the distal end, abscission is accelerated. Under these conditions, ABA acts to enhance the rate of senescence in the distal tissue and as a result, the senescence-associated ethylene production of the distal tissue occurs sooner. The initiation of the abscission process is therefore brought forward in time in response to the earlier ethylene production.[13] Confirmation that the control of abscission lies with ethylene rather than directly with ABA is given by experiments in which ethylene production is blocked by the addition of an ethylene biosynthesis inhibitor such as α-aminooxyacetic acid (AOA) or 1-aminoethoxyvinylglycine (AVG). In the presence of these inhibitors, ABA no longer exerts an abscission-accelerating action though abscission can still be induced to occur if the explants are exposed to ethylene, showing that the ability to separate has not been impaired by the inhibitor treatments.

The difference in zone cell recognition and response between dicotyledons and the Graminae has implications for agriculture; perhaps the most important is the opportunity it offers to induce a differential shedding response in the two groups of plants by the foliar application of either ethylene-releasing compounds or abscisic acid treatments. The separation of both kinds of zones is, however, repressed by auxin or auxin analogs so it is only the signal for initiating abscission in these two types of target cells that is different.

III. DIFFERENTIATION OF TARGET CELLS

Every plant cell is in practice a captive of its neighbors. From them and through them it receives local cues and long-distance internal information which will include hormones, nutrients, and metabolites. Imposed on all the cells are also the external environmental signals of light, water availability, and temperature.

On the way to a final developmental state in the mature plant a cell will pass through a progression of different target states, influenced in this progression by both internal and external signals which it may or may not be able to recognize depending upon its condition of competence at the time.

The sequential changes that take place represent a series of modifications in gene expression in which the signal recognition and response of the cell is progressively modified in a highly precise and specific manner.

As each new signal recognition and response is achieved and as ecah new state of competence is superseded by another, the options remaining to a cell for further differentiation and development become progressively restricted.

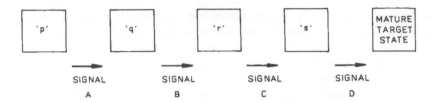

FIGURE 2. A program of sequential signal recognitions and responses during cell differentiation.

A. Developmental Programs at the Cell Level

We can envisage (Figure 2) a cell responding to a series of internal or external signals during its growth and development to the mature target state. For example, cell type 'p' after it receives and responds to Signal A can differentiate to cell type 'q', and so on.

Each cell type has a cell-specific (or in a collection of cells, a tissue-specific) phenotypic expression which we should — if we looked — recognize biochemically by a cell-specific protein complement. The cell-specific protein complement, or more precisely, the specific expression of the genome, gives the cell its target status — in the case of cell type 'q' it signifies the potential competence to recognize Signal B and progress to cell type 'r'. On this model, cell type 'p' cannot recognize Signal B and cannot therefore differentiate to cell type 'r'. However, it must be assumed that cell type 'p' has the potential to recognize a number of other signals besides Signal A. Let us suppose it can recognize Signals X, Y, or Z. The cell would then progress along a different set of developmental signal recognitions and responses — but it cannot then progress through the phase transitions of the cell types that eventually become the same mature target cell derived through the 'p','q', 'r's pathway.

B. Conditions of Cell Recognition and Response
1. The Stage of the Cell Cycle

The cells of an actively dividing meristem will encounter internal and external signals throughout the cell cycle. However, if a signal can be perceived and responded to only at a critical stage in the cell cycle then the "time slot" available for recognition leading to response may be short.

The activation of the mitotic cycle in the shoot meristem of *Sinapis alba* following photoinduction or cytokinin treatment may therefore be interpreted as an "enabling" situation that maximizes the number of cells that will be in the right state of the cell cycle to respond to a flowering signal.[9] Critical light or dark temporal recognitions of hormone signals occur in *Acetabularia* in which auxin will accelerate cap formation when present in the early, but not the later, part of the dark phase and where ethylene control is differentially effective with time of day.[14]

Such short-term temporal "recognition slots" imply that although the chemical signal, e.g., Signal A, may be present in a tissue in "response-controlling" concentrations over a long period, recognition and response could be rapid events within the cell thereby changing its developmental target status from, e.g., cell type 'p' to cell type 's' without any change in the measurable level of Signal A in the cell. We know from experiments with soybean seedlings that treatment with auxin can elicit the production of elongation-specific polypeptides and suppress the production of others within 1 hr. Auxin-induced change in in vitro translatable mRNA populations in both elongating soybean and pea epicotyls occur even sooner, within 15 or 20 min, respectively.[15]

2. Cell Age

The performance of a specific cell type may also be modified with age. The basis of age-related change, particularly with respect to hormonal responses, is presently little understood.

In tobacco pith cultures, for example, habituation for cytokinin synthesis is greater in those cultures derived from pith excised from the apical parts of stems rather than from the older basal parts. This raises the question of whether the target status and the gene expression of these two differently aged, but apparently similar, cell types is, in fact, quite the same and serves to emphasize the importance of the effect of the simple difference of age on the extent of the response that can be expected from the application of any chemical substance to the plant foliage.

IV. TARGET CELLS AND PHYSIOLOGICAL RESPONSES

A. Cell-Specific Proteins

The concept that each target state of a cell is marked by a particular complement of proteins diagnostic for that specific committed condition raises questions on the biochemical mechanisms for recognition of internal and external signals such that a cell-specific response is elicited. Some consider that the specific protein population (or mRNA abundancies that direct protein pattern) include those for hormone-binding proteins or specific receptors which will operate for a variety of small molecules with growth-regulatory activity. It should be said that although many proteins showing a high-affinity binding for growth-regulating molecules have now been extracted from plants, none has yet risen to the status of "receptor" for, so far, binding has not been unequivocally shown to be prerequisite to a demonstrable hormonally controlled physiological response.[16] Because cell surface receptor proteins are clearly involved in the hormone-mediated changes in mammalian cells (such as insulin in liver and steroids in uterine tissue)[17] there has been keen research to discover if similar systems operate in plants. However, a comparably functioning recognition system may not exist! For example, it may be that all plant cells contain auxin- (and other hormone) binding sites located at the plasma membrane and these binding sites could be "carrier" proteins acting solely in the transfer of hormones across the plasma membrane.[18] Such carriers could be nonlimiting to uptake and so function differently from the surface membrane receptors in animals where the entry of the hormone into the cell is regulated by the abundancy of the receptor molecules at the membrane surface. The effective hormone (or hormone analog) "receptors" in the plant cell may be wholly cytoplasmic. Many different receptors could be present at very low concentrations only, resulting in a potential plurality of possible responses; the target response being determined by the receptor currently in greatest abundance.

Depending upon the temporal gene expression, the protein complement, and the receptor availability a cell can be seen as exhibiting a number of options in terms of its physiological response. There is no need to invoke very many receptors to accommodate the recognition of a diversity of signals. Referring again to Figure 2 it can be envisaged that cell type 'p' may recognize and respond to the same signal in its passage to 'q' as cell type 's' recognizes and responds to on its transition to the mature target state. The response of the cell to any internal or external signal is therefore ultimately determined by the protein complement which reflects the molecular competence of the specific differentiated target state.[4]

B. Sensitivity

In recent years there has been some discussion of the "sensitivity" of a plant tissue to an applied hormone (or other chemical) and this term has been used to explain differential responses of cells to external or internal stimuli.[19] Interpreted in the target-cell concept, this more precisely means that a cell may show a *quantitatively* regulated response to the signal it perceives. Certainly the change from cell type 'p' to cell type 'q' cannot be instantaneous so there will be a period in which gene transcripts much reach a threshold level for the elicitation of the new target-cell response. Once the competence of that target type is attained, however, the cell is fully sensitive and remains fully sensitive as long as it retains that specific target state. The degree of response to its target signal depends solely upon the intensity of the signal and the presence of inhibitors or derepressors that can block the signal recognition or response. Taken overall, therefore, the plant and biochemical constitution of its individual tissues and differential target competence of its cells determines the responses that can be evoked when any kind of chemical is applied. It is a challenge for the scientist to discover how a plant perceives and translates these external signals into the physiological responses that we eventually observe.

V. IMPLICATIONS FOR PLANT RESPONSES TO LEAF-APPLIED CHEMICALS

When a chemical is applied to a plant an array of different factors assumes significance in determing the eventual morphological and biochemical changes that occur. Not the least of these is the target status of every cell, its condition with respect to the differentiation of its neighbors, its stage in the cell cycle, its age, and its past history.

It is small wonder that coupled with the vagaries of environmental cues, the result of a foliage treatment can give a wide spectrum of results. The more uniform a crop the easier it is to predict the outcome. In broad terms we get it right, and experience and experimentation tell us what to do and what not to do. However, failures of plant manipulation and unexpected crop losses are the starting points for our better understanding for they provide the clues and the background for the next advances in controlling plants.

ACKNOWLEDGMENTS

I wish to thank Miss Claire Gurney and Miss Janice Spruce for their valuable assistance with the preparation of this manuscript.

REFERENCES

1. **Slade, R. E., Templeman, W. G., and Sexton, W. A.,** Plant growth substances as selective weed-killers, *Nature (London),* 153, 407, 1945.
2. **Moreland, D. E.,** Mechanism of action of herbicides, *Annu. Rev. Plant Physiol.,* 31, 597, 1980.
3. **Osborne, D. J.,** Concepts of target cells in plant differentation, *Cell Differ.,* 14,161, 1984.
4. **Osborne, D. J. and McManus, M. T.,** Flexibility and commitment in plant cells during development, *Curr. Top. Dev. Biol.,* 20, 383, 1986.
5. **Osborne, D. J.,** The ethylene regulation of cell growth in specific target tissues of plants, in *Plant Growth Substances 1982,* Wareing, P. F., Ed., Academic Press, London, 1982, 279.
6. **Taiz, L.,** Plant cell expansion: regulation of cell wall mechanical properties, *Annu. Rev. Plant Physiol.,* 35, 585, 1984.
7. **Osborne, D. J., McManus, M. T., and Webb, J.,** Target cells for ethylene action, in *Ethylene and Plant Development,* Roberts, J. A. and Tucker, G. A., Eds., Butterworths, London, 1985, 197.
8. **Osborne, D. J.,** Ethylene and plants of aquatic and semi-aquatic environments, *Plant Growth Regul.,* 2, 167, 1984.
9. **Pierard, D., Jacqmard, A., Bernier, G., and Salmon, J.,** Appearance and disappearance of proteins in the shoot apical meristem of *Sinapis alba* in transition to flowering, *Planta,* 150, 397, 1980.
10. **Rascio, N., Ramina, A., Masia, A., and Carlotti, C.,** Leaf abscission in peach *(Prunus persica* L. Batsch): ultrastructural and biochemical aspects, *Planta,* in press, 1987.
11. **Sargent, J. A., Osborne, D. J., and Dunford, S. M.,** Cell separation and its hormonal control during fruit abscission in the Graminae, *J. Exp. Bot.,* 35, 1663, 1984.
12. **Osborne, D. J.,** Abscission in agriculture, *Outlook Agric.,* 13, 97, 1984.
13. **Jackson, M. B. and Osborne, D. J.,** Abscisic acid, auxin, and ethylene in explant abscission, *J. Exp. Bot.,* 23, 849, 1972.
14. **Vanden Driessche, T.,** Temporal morphology and cap formation in *Acetabularia.* II. Effects of morphactin and auxin, *Chronobiol. Int.,* 1, 113, 1984.

15. **Theologis, A.,** Rapid gene regulation by auxin, *Annu. Rev. Plant Physiol.,* 37, 407, 1986.
16. **Firn, R. D., and Kearns, A. W.,** The search for the auxin receptor, in *Plant Growth Substances 1982,* Wareing, P. F., Ed., Academic Press, London, 1982, 385.
17. **Cuatrecasas, P. and Hollenberg, M. D.,** Membrane receptors and hormone action, *Adv. Protein Chem.,* 30, 251, 1976.
18. **Rubery, P. H.,** Auxin receptors, *Annu. Rev. Plant Physiol.,* 32, 569, 1981.
19. **Trewavas, A. J.,** How do plant growth substances work?, *Plant Cell Environ.,* 4, 203, 1981.

Chapter 5

FOLIAR NUTRITION OF FRUIT TREES

Steven A. Weinbaum

TABLE OF CONTENTS

I. INTRODUCTION AND SCOPE OF REVIEW

Foliar nutrition of fruit crops has been discussed recently,[38,108] and interested readers are referred to those excellent and broad-based reviews. Our intention is not to review the whole subject as limited developments during the brief intervening period do not warrant that effort. Rather, our intent is to develop a quantitative perspective from which to discuss and evaluate the advantages and limitations of foliage application of nutrients in fruit trees. As the effectiveness of foliage-applied nutrients often depends on their redistribution within the plant from the leaves treated, factors influencing nutrient redistribution will also be addressed.

Foliage application of plant growth-regulating chemicals (PGRs) used by tree fruit growers around the world is unmatched in commercial agriculture. PGRs are used in the production of nursery trees, also, to regulate the shape and size of orchard trees, promote flowering, and increase and reduce fruit set. Other treatments inhibit the growth of unwanted shoots or suckers, and still others prevent preharvest fruit drop, regulate fruit ripening, and influence various aspects of fruit quality. These subjects have been reviewed by Looney[72] and are outside the scope of the present discussion. The management of insect and disease pests in tree fruit crops depends heavily on foliage application of pesticides, but these areas are also beyond the focus of the present discussion

A quantitative perspective is clearly essential if we are to appreciate the limitations of foliage-applied nutrition as well as its potential. For example, the desirability of minimizing ground water contamination resulting from excessive rates and/or application strategies of soil-applied nitrogenous fertilizers to agricultural soils is apparent. It has been proposed that foliage application of nutrients would minimize these dangers.[25,52] Yet, the nutrient demand of fruit trees must, nevertheless, be met if the solution proposed by environmental interests is to be compatible with the answer acceptable to commercial agriculture.

The potential significance of foliage-applied nutrition relative to the total nutrient demand of the tree would appear to be influenced by at least six factors:

1. The timing and level of nutrient demand with respect to the presence of the leaf canopy
2. Canopy size and persistence, i.e., the size of the potential absorptive surface, and the life span of the foliage
3. The concentration of nutrient(s) in solution which may be applied to the foliage without serious phytotoxicity
4. The degree of leaf wetting, cuticular penetration, and leaf absorption following foliage application
5. Nutrient redistribution (phloem mobility) from the leaves to sites of nutrient demand within the tree
6. Circumstances resulting in the inability of trees to acquire sufficient quantities of nutrients from the soil to achieve optimal growth and yield

The relative demand by plants for essential nutrients[2] is expressed by the macro- or micronutrient status of essential plant nutrients. Macroelements, e.g., nitrogen, phosphorus, potassium, etc., are present in leaves at levels between 0.1 to 10% of leaf dry weight. Micronutrients or trace elements (iron, boron, manganese, copper, zinc, and molybdenum) are required at much lower concentrations, i.e., 0.0001 to 0.01% of leaf dry weight.

Discussions of foliage application of nutrients invariably are limited to elements known to be essential for completion of the plant life cycle. The recent focus on the beneficial role of foliage-applied titanium, an element not presently considered essential, is intriguing.[89] Pais[89] reported yield increases in apple, peach, apricot, gooseberry, and black currant in response to the application of a water-soluble titanium chelate.

II. RELATIONSHIP BETWEEN PLANT NUTRIENT STATUS AND RESPONSE TO FOLIAGE APPLICATION OF NUTRIENTS

The orchardman is exposed continually to commercial interests touting (without reservation and adequate supporting data) the advantages of foliage application of nutrients. The rapidity and efficiency of nutrient recovery by the tree following foliage application is well documented.[18] The economic advantage of foliage nutrient applications in the absence of diagnosed mineral deficiencies, however, has received relatively little critical attention.

Oland[84-87] in Norway reported that foliage application of 4% urea to apple in autumn increased the amount of nitrogen remobilization from leaves during senescence, which resulted in a 31% increase in nitrogen concentration of reproductive spurs by leaf fall (November 12), and increased fruit yield the subsequent year. Other attempts to reproduce his results have failed, however.[71,123] This discrepancy may be associated with differences in tree nitrogen status, since the 'Gravenstein' trees on which Oland obtained a response had only 1.5 to 2.0% dry weight nitrogen in their leaves when sampled in September, in contrast to the more typical value of 2.4% dry weight nitrogen in similar samples from 'Gravenstein' trees of adequate nitrogen status. It may be concluded that foliage application of urea is beneficial when nitrogen deficiency exists, but it is unlikely that foliage application of urea in autumn would affect cropping in the average, well-managed orchard.[22] Conditions which potentially limit the availability of nutrients in the soil or their utilization (e.g., shallow, light-textured soils, late fruit harvests, early autumn frosts, late thawing of soils in spring, etc.) represent situations in which foliage application of nutrients offers an important alternative means of supply for maintaining tree nutrient status.

Multiple applications of ^{15}N-enriched 2% urea in October to the foliage of olive trees of adequate nitrogen status increased leaf nitrogen significantly, but did not affect the percentage of flowering nodes or flower size at anthesis in May.[57] Flower nitrogen concentration was similarly unaffected, although 4.3% of the nitrogen in the inflorescence in May was derived from applications of urea the previous autumn. Similarly, foliage application of ^{15}N-enriched urea to almond trees in October (applied as a 2% solution) resulted in remobilization of labeled nitrogen from the senescing leaves to storage pools in the perennial parts of the trees. Labeled nitrogen derived from foliage-applied urea the previous autumn accounted for about 3% of the blossom nitrogen at anthesis in February, but the treatment did not affect flower size, nitrogen concentration, or fruit set (Weinbaum, unpublished).

It appears, therefore, that an inverse relationship between tree nutrient status and the likelihood of response to foliage application of the nutrient in question may be anticipated. Foliage application of nutrients in the absence of incipient nutrient deficiency, thus, appears to be of questionable value.

This concept may be less valid when considering the status of the tree with respect to nutrients less mobile in the phloem (e.g., calcium, boron, and iron, see later discussion).

In these instances, the nutrient may be available in the soil and within the plant (including the leaves), but because of limited phloem mobility, the nutrient is not available to reproductive tissues. These cases may be described as reproductive tissue deficiency rather than plant deficiency, and the effectiveness of "foliage" application would appear to be dependent upon the direct application of the deficient nutrient to the flower/fruit rather than the leaves.

III. RELATIONSHIP BETWEEN TIMING AND EFFECTIVENESS OF FOLIAGE APPLICATION OF NUTRIENTS

Possible advantages of foliage nutrient applications would appear associated with periods and/or situations in which the capacity for natural acquisition of nutrients by fruit trees is reduced. Foliage applications of nutrients are typically made (1) in early spring to increase the percentage fruit set, (2) during the growing season in response to the immediate need for a deficient nutrient, and (3) as postharvest applications in summer/autumn to augment nutrient reserves in the branches available to support the spring flush of vegetative and reproductive growth[37,84-87] occurring before the roots are capable of significant uptake of soil nutrients.[116,118] Dormant applications of zinc, practiced as a consequence of early concerns about plant injury from foliar $ZnSO_4$ applications,[79] are less satisfactory or practical than foliage sprays. Foliar applications in autumn of 5 to 10 lb of zinc sulfate (36% zinc) per 100 Gal (20 to 40 lb/acre) give good correction if made in October before leaf drop. Sprays will give correction in almonds, apricots, cherries, and pears. Some burn injury can occur on peaches so the lower rate is maximum.[6]

In early spring, the carbon and nutrient resources which support the initial phases of the spring flush of growth in deciduous fruit trees are not derived from concurrent carbon assimilation and nutrient uptake. Rather, organic metabolic reserves and inorganic nutrients[37,109] are accumulated in late summer/autumn in the perennial vegetative structures (roots, trunk, and branches) and are redistributed from these storage pools to support the spring flush of growth. In evergreen fruit trees (e.g., *Citrus* spp. and olive), leaves represent major reservoirs of nutrients which are similarly utilized in spring.[13,77,57,92] The availability of nutrients from storage thus represents a nutrient buffer until soil conditions favor nutrient absorption by roots.

Although foliage-applied urea has not increased yield in apple trees judged to be adequate in nitrogen status, the reported stimulation of initial fruit set by foliage application of urea[101,125] is intriguing. A pulse of quickly available foliage-applied nitrogen at bloom may correct a transient deficiency of nitrogen which could conceivably occur between exhaustion of nitrogen from within-tree storage pools and significant uptake of nitrogen by roots in spring. Stimulation of fruit set by urea sprays in spring may be more effective in immature than mature fruit trees because the buffering capacity of the endogenous pool of storage nitrogen is reduced in the former.[41] Nevertheless, the quantity of nutrient potentially absorbed by deciduous fruit trees following foliage application in early spring would appear to be restricted by the limited leaf area.

The presumed advantages of foliage, as compared with soil, application of nutrients are related to limitations in the availability of nutrients in the soil coupled with the immediate demand of the developing fruit and foliage for the growth-limiting nutrient. The immediate availability of nutrients supplied via the foliage[85] (1 to 2 days) and rapid plant response[28] as compared with soil application is typical.

Limitations in soil nutrient availability may have both direct and indirect causes. Direct limitations include soil nutrient deficiency or immobilization of nutrients in the soil. Potassium fixation in many fine-textured soils in California results in serious potassium deficiency expressed as leaf scorching and shoot die-back as well as sunburned and undersized fruit of heavily cropping French prune *(Prunus domestica)* trees.[70,113] Similar symptoms have been reported in heavily cropping trees of pecan *(Carya illinoensis)* in the southeastern U.S.[106] Foliage application of potassium in prune[12] and iron in citrus[45] has proven effective in instances when massive soil applications (e.g., 1500 to 2500 kg K_2SO_4 per hectare) have been ineffectual and/or not economical.[113] Fruit are sinks for potassium, and potassium is mobilized by fruit whether or not there is an adequate supply for the leaves. The problem is presumably exaggerated by the adverse effect of the heavy crop load on the availability and transport of assimilates to roots.[12] This results in reduced root growth (exploration of the soil volume) and capacity for nutrient acquisition. The only means available to alleviate symptoms in the season they appear is through the use of foliage applications of potassium (supplied routinely as KNO_3). Two to three sprays 2 weeks apart (30 lb/acre in 100 to 400 Gal prevent severe limb defoliation and dieback (O. Lilleland, unpublished).

Soil environmental parameters (principally soil moisture, soil temperature, and soil oxygen tension) can influence nutrient uptake and translocation by roots. Root temperature optima have been shown for nutrient uptake and assimilation.[4,110] The temperature optimum for potassium uptake by roots of appled trees occurred at about 20°C, and uptake was reduced significantly at temperatures below 6°C and supraoptimal temperatures above 24°C.[110,34] Foliage application of potassium salts to apple trees experiencing supraoptimal root temperatures of 29°C and 36°C reportedly alleviated symptoms at 29°C but not at 36°C.[35] In general, however, the potential of foliar nutrient applications to alleviate nutrient deficiency associated with adverse soil temperatures is not well documented. Root temperature optima have been shown to vary among rootstock clones.[35,110] As soil temperatures measured within the root zone (50 cm in depth) in the Mediterranean climate at Davis, California, U.S. appear suboptimal, nutrient uptake by roots may be limited by low temperature during the early spring growth flush, i.e., prior to April.[104] In wetter, heavier-textured soils or in colder climates greater limitations to nutrient uptake by roots in spring resulting from suboptimal root temperatures may be anticipated. Differences in soil temperature could also influence foliage content of essential elements by modifying the availability of the elements in the soil due to differences in microbial activity or by the size and extent of the root system. With elements which remain immobile in the soil, such as phosphorus, the roots must grow to the ion. Unfavorable soil temperatures which reduce root growth may result in nutrient deficiencies.[4] It is known that at the lower root temperatures which are characteristic during anthesis in apple, nutrient translocation from the roots is restricted severely.[50] Suboptimal root temperatures not only reduce nutrient acquisition and transport[35,110] but also water absorption[4,60] and the synthesis and transport from roots of endogenous hormones such as cytokinins and gibberellins having shoot-regulatory activity.[50] Supraoptimal root temperatures (which presumably occur only rarely at soil depths >10 cm) may not only reduce nutrient uptake and the biosynthesis and transport of endogenous hormones, but present the affected plant with additional challenges such as the accumulation of potentially toxic metabolites.[34]

Exposure of apple roots to supraoptimal root temperatures may create conditions similar to those resulting from low soil oxygen supply (e.g., 2% O_2) brought about by waterlogging.[35,50,61] At field capacity, the limiting oxygen pressure at the root surface may correspond to a requirement for 15 to 25% airfilled pore space or at least 10 to 15% oxygen in the soil atmosphere.[121] Thus, if plants experience root anoxia brought about either by supraoptimal root temperatures or waterlogging, the problem is considerably more complicated than making good any deficiencies arising from inadequate inorganic nutrient uptake. Inhibiting the effects of accumulated gases, notably ethylene, overcoming the action of potentially toxic products of anaerobic metabolism, and the correction of hormonal imbalances would all appear necessary to permit normal plant development.[50] Attempts to rectify the hormone imbalance in tomato by foliage application of GAs and cytokinins resulted in only a transient reversal of waterlogging injury.[50] Thus, overcoming the effects of waterlogging by means of foliage application of chemicals has not yet been achieved.[94]

Nutrient absorption by roots is also decreased in plants under water stress[111] and foliage application of nutrients offers the possibility of an alternative path for nutrient entry.[66] Root factors such as morphology and extension growth may be involved, but, in addition, the translocation of nutrients in the soil toward the root surface is known to be severely impeded under drought conditions and of far greater significance than the metabolic capacity of roots for absorption.[96,120]

The effectiveness of supplementing the nutrient reserves of fruit trees by postharvest foliage applications of nutrients, particularly urea, is well documented.[42,85] The quantitative advantage associated with foliage application of nutrients during the postharvest period is based on the following parameters:

1. Timing — postharvest applications coincide with the natural remobilization of phloem-

mobile leaf nutrients by the tree. Leaf abscission which occurred even within a few weeks of foliage application of nitrogen was preceded by translocation of half the foliage-applied nitrogen back into the tree.[42]

2.   Leaf phytotoxicity — uptake of foliage-applied nutrients is reportedly proportional to the quantity of nutrient applied (number of sprays times nutrient concentration).[88]

The chemical concentration which may be applied during most of the growing season, however, is limited by the concentration resulting in leaf phytotoxicity. During the postharvest period, however, particularly as the period of natural leaf fall approaches, there is less concern with retaining the functional integrity of the foliage and therefore, the concentration of the nutrient in the solution applied may be increased much beyond levels tolerated during the growing season.

The advantages associated with foliage application of nutrients during the postharvest period to supplement nutrient storage pools within the tree should be limited to the use of phloem-mobile nutrients. Thus, the accumulation of phloem-immobile elements, particularly calcium, in senescent leaf tissue would seem to preclude any advantage of supplying trees with this element during the postharvest period.[87] Boron appears to be an anomaly in this respect. Although boron is considered phloem immobile,[108] foliage application of boron in fall increased the percentage fruit set in prune flowers the subsequent spring.[37] Hansen and Breen[37] reported that the percentage redistribution of foliage-applied boron from leaves in autumn was about three times greater ($\cong 70$ vs. 25%) than the percentage of total leaf boron remobilized from untreated leaves prior to leaf senescence and abscission. These authors suggested that foliage-applied boron may enrich a pool of boron in the leaf which is soluble and highly mobile.

In assessing the quantitative significance of nutrient remobilization from senescing leaves to the perennial vegetative framework, it should not be assumed that nutrient disappearance from leaves is necessarily indicative of nutrient remobilization. Nutrients such as potassium may be lost from foliage by leaching.[21,73] This mechanism may account for >50% loss of leaf potassium.[18] Also, significant loss of nitrogen volatiles may occur from leaves in conjunction with the loss of water vapor during transpiration.[102,115]

IV. "EFFICIENCY" OF FOLIAGE APPLICATION OF NUTRIENTS

The "efficiency" of foliage nutrient applications may be considered from several perspectives: (1) the rapidity of nutrient availability following foliage application, (2) the percentage absorption of foliage-applied nutrients, (3) the percentage absorption of foliage-applied nutrients as compared with the percentage recovery following application of nutrients to soils, and (4) the actual amounts of foliage-applied nutrients absorbed. With respect to the last point, we may ask how effectively can foliage application meet the nutrient demands of the tree. Clearly, demand and the potential for significant nutrient supply via foliage varies among essential nutrients, and is influenced further by the nutrient status of the plant, tree vigor, and cropping intensity. The relative significance of a foliage application of a macroelement such as nitrogen is indicated by the fact that it increased the endogenous nitrogen content of treated leaves of European plum *(Prunus domestica* L.) by only 3%.[117] In contrast, foliage application of a microelement (boron to *Prunus domestica* L.) resulted in a 300% increase in the boron concentration of the leaves treated.[37] The quantity of nutrient which must be replaced annually in a mature tree may be approximated conservatively (i.e., does not include losses associated with pruning, root turnover, and vegetative increment) by the amount of nutrient removed in the crop plus a percentage of nutrients lost in the leaf litter following leaf abscission.

Any economic assessment of foliage application of nutrients must recognize limitations (due to phytotoxicity) in the concentration and, therefore, nutrient amounts that can be applied in one spray. These limitations are particularly serious when considering macroelements (particularly potassium and nitrogen). Several sprays a year may be required. Although one annual application of soil nitrogen provided sufficient nitrogen for citrus trees, six or more foliar sprays were required.[25] Because more application-equipment traffic is associated with foliage, as compared to soil-applied nitrogen, the potential for greater soil compaction exists.[25] It is obvious that application costs (fuel, labor, and other operating costs) are greater for foliar than for soil-applied nitrogen.[25] These factors should be considered in any economic evaluation of the commercial implications of foliage application of nutrients. Nevertheless, if the fertilizer salts are compatible in the spray tank with the normal insecticide and fungicide spray applications, there is no additional economic burden associated with the foliage application of nutrients.

Foliage applications of nitrogen and potassium typically result in more rapid uptake, but responses are more transient as compared with soil applications.[28] The very high efficiency of foliage-applied nitrogen (using urea as the carrier) is well documented and, 80% uptake of foliage-applied urea within 24 to 48 hr has been reported in a number of tree fruit species.[57,58]

The percentage recovery of foliage-applied nitrogen in the tree is reportedly three to four times higher than the recovery following soil application.[101] A comparison of fertilizer recovery by potted apple trees following soil or foliage-applied labeled nitrogen in October indicates a much higher percentage recovery (70 vs. 16%) associated with foliar application.[42] The extent of this advantage would be reduced if the comparison was made during periods of active tree growth, e.g., March.[42] Although soil application of 450 g nitrogen per tree to prune trees (*Prunus domestica* L.) increased leaf nitrogen concentration and fruit yield, foliage application of 75 g nitrogen per tree (applied as $4 \times 0.5\%$ urea) neither increased yield nor resulted in a measurable increase in leaf nitrogen concentration.[65] Postharvest foliage application of 400 g nitrogen per tree increased leaf nitrogen concentration by 15%.[65]

Deciduous fruit tree species appear relatively inefficient in their extraction of soil applied nitrogen.[20,75] Using ^{15}N-depleted $(NH_4)_2SO_4$, we have determined by direct measurement that annual recovery of soil-applied fertilizer nitrogen in the crop of mature almond trees is <20%.[118] Annual recovery of all soil sources of nitrogen by the tree is more difficult to determine but our data (Weinbaum and Klein, manuscript in preparation) indicate >50% recovery. The availability of soil-applied nutrients varies greatly and is affected by biological and chemical immobilization, i.e., soil texture and microorganisms, soil pH, leaching, soil temperature, soil aeration, soil moisture, etc.

Efficiency, i.e., percentage recovery of the applied nutrient, must be distinguished from the capacity for supply. That is, utilization of foliage-applied nutrients may be highly efficient (e.g., 80%) but grossly inadequate to meet the magnitude of plant demand. A quantitative perspective, although virtually ignored in the horticultural literature, may be developed by coordinating data from a variety of sources. For example, using foliage-applied ^{15}N-enriched KNO_3 (1.2%) we have determined that a single application could provide 10.5 g of nitrogen per mature prune tree.[117] Based on average yields of French prune in California and nitrogen concentration of the harvested fruit (K. Uriu, personal communication) we have calculated the removal of about 208 g nitrogen per tree by the crop. Thus, a single foliage application of KNO_3, applied at the limits of phytotoxicity,[116] resulted in nitrogen absorption equivalent to only about 5% of the nitrogen removed in the crop (note: not total tree demand). Weinbaum[116] estimated that a single application of foliage-delivered NO_3^- could provide only 0.69% of the total seasonal nitrogen demand of 2-year old nonbearing prune trees. With this perspective, the inability of six foliage applications of urea to maintain apple tree vigor[29] is not surprising even if urea (45% nitrogen) is a more efficient nitrogen source than NO_3^- (14% nitrogen).

Similarly, in French prune, the 0.2% increase in potassium concentration in prune leaves following a KNO$_3$ application (10 lb 100 Gal^{-1}) (K. Uriu, personal communication) average leaf dry weight (180 mg), and average leaf number per tree (100,000; O. Lilleland, personal communication) can be used to estimate a potassium contribution of 36 g tree^{-1} resulting from a single foliage application of KNO$_3$. Estimated yearly demand by normally cropping mature prune trees is 590 g potassium tree^{-1}.[12] A simple calculation indicates that only 6% of the estimated annual demand for potassium by mature prune trees can be met by a single foliage application of KNO$_3$. Because of the magnitude of need, it appears unlikely that foliage application of macronutrients in deciduous fruit trees could replace completely the need for soil-derived nutrients.

Development of comparable quantitative estimates of the level of micronutrient demand which may be satisfied by foliage application are plagued by the occurrence of large relative errors associated with cuticular adsorption and, as a result, incomplete removal of micronutrients (e.g., zinc) from leaf surfaces even following acid washing prior to analysis (R. M. Carlson, personal communication). Also, limited phloem mobility and, therefore, minimal redistribution of nutrients from treated leaves complicate the situation.

V. DIFFERENCES AMONG SPECIES IN THE EFFECTIVENESS OF FOLIAGE APPLICATION OF COMPOUNDS

Citrus spp. represent the only instances for which there is documentation that urea sprays are able to replace completely the need for soil nitrogen applications.[25] Fruit species may differ greatly in response to some foliage-applied chemicals, especially urea, and *Prunus* spp. (the stone fruits) were found to be much less responsive than *Citrus*.[63] The main difference among species may reflect differential sensitivity of the foliage, i.e., phytotoxicity or differences in cuticular penetration. Cuticular penetration is reportedly the rate-limiting step during absorption of foliage-applied nutrients.[68] The wax content of the abaxial surface of the leaf was the cuticle constituent that was best correlated with the differential responses of peach, apple, and orange to foliar sprays.[63]

Besides a more efficient uptake of foliage-applied nutrients, it has been suggested that the evergreens, such as citrus, are more efficient in their utilization and reutilization of nutrients because of their longer leaf life spans which moderate the recycling of nutrients through the litter fall and decomposition process.[13] Thus, the annual demand for nutrient acquisition in evergreens may be lower. In this situation, the contribution of foliage-applied nutrients to total nutrient demand may be quantitatively more significant.[25]

Uptake of foliage-applied nutrients is proportional to the concentration of nutrient applied,[9,58,85,88] but the potential for increasing the concentration of nutrients in the solutions applied is limited by the tolerance of the foliage.[67,83] This tolerance varies among nutrient carriers, leaf phenology, and species. Thus, e.g., in the early postbloom period (May) phytotoxicity results when apple trees receive foliage applications of urea at concentrations >0.5%. In contrast, 4 to 5% urea solutions may be applied to the same trees in October with greatly reduced phytotoxicity. Also, some injury to the leaves shortly before leaf fall is unlikely to affect adversely tree productivity[85-87] (see also references cited in Weinbaum[116]). Unpublished evidence indicates, however, that leaf damage reduces the efficiency of uptake and redistribution of foliage-applied urea (I. Klein, personal communication).

The relative contribution, and thus, significance, of foliage-applied nutrients may also vary greatly among species. Thus, when the highest concentration of urea that can be used without causing phytotoxicity (0.5% for almond and 4% for olive) is considered, we have calculated that a single foliage application will deposit approximately 15 times more urea per square centimeter of an olive leaf than per square centimeter of an almond leaf. The 15 times greater deposition of urea per square centimeter leaf area on olive leaves vs. almond

Table 1
INFLUENCE OF TREE GROWTH AND BEARING HABIT
(I.E., LONG SHOOT — 'ELBERTA' PEACH OR SPUR
SHOOT — 'McINTOSH' APPLE) ON SEASONAL
DEVELOPMENT OF LEAF CANOPY

Percentage completion[a] of growing season	Leaf canopy development (Expressed as % of maximum canopy attained)	
	'Elberta' peach[b]	'McIntosh' apple[c]
10	—[d]	36
20	—	74
30	—	89
40	—	92
50	50	97
60	65	100
70	78	100
80	90	100
90	98	100
100	100	100

[a] Growing season computations based on the period from full bloom to fruit harvest.
[b] Calculations based on data of Ryugo.[95]
[c] Calculations based on data of Lakso.[62]
[d] Data not collected.

leaves, rather than the 8 times difference anticipated on the basis of the concentrations applied (5 vs. 0.5%) is due presumably to differences between the species in surface wetting properties.[44,48]

Price and Anderson[91] evaluated the uptake from foliar deposits of ten ^{14}C-labeled compounds into field-grown plants. These compounds were selected as representative of pesticide and plant growth regulators. Apple and orange, the only tree fruit species among the ten species tested, exhibited the poorest uptake, but the available data on the structure, composition, and thickness of the cuticles and epicuticular waxes of the species tested are insufficient to be able to relate species differences to patterns of uptake. These authors observed, however, that both apple and orange have particularly thick cuticles (about 1 to 1.5 μM).

VI. LEAF CANOPY DEVELOPMENT, LEAF HOMOGENEITY, AND LEAF PERSISTENCE ON EFFICIENCY OF FOLIAGE APPLICATION OF NUTRIENTS

The dynamics of canopy development (total canopy size, length of growing season, and persistence of the canopy) i.e., the size and temporal limits of the absorptive surface influence the efficiency of foliage application of chemicals.

Differences among species in bearing habit as well as tree age may affect the homogeneity of the leaf canopy. In many tree fruit species there are two distinct classes of shoots; long shoots and short shoots (also called spur shoots; typically 1 to 10 cm long with very short internodes). In many species, in fact, there is a continuous gradation in shoot length between the two shoot types.[122] In peach *(Prunus persica)* the canopy is borne entirely on long shoots, and shoot growth, leaf initiation, and expansion may continue throughout the growing season.[95] In contrast (Table 1), mature apple *(Malus domestica)* trees bear their fruit and canopy primarily on spur shoots, and development of the apple leaf canopy of mature apple trees is essentially complete within a month.[62] The canopy of the almond *(Prunus dulcis)* is borne both on spur shoots and long shoots, and the ratio of spur shoots to long shoots increases as almond trees become older and less vigorous.[16] The relevance of shoot type to the present discussion involves its correlation with rate of canopy development, canopy persistence, and the distribution of ontogenetic leaf ages within the canopy. Typically, the entire complement of spur leaves is preformed and, following budbreak, these early leaves expand rapidly. Late leaf initiation and expansion in the current year accompanies the development of long shoots on young vigorous trees later into the season.[122]

The onset and rate of canopy development may also be influenced by the degree to which the chilling requirement of deciduous fruit trees and/or the accumulation of heat units has been met.[16,17] Insufficient winter chilling results in both delayed budbreak and a slower rate of leaf canopy development following budbreak.

The persistence of the leaf canopy, i.e., duration of the absorptive surface, has obvious significance with respect to the potential for foliage application and absorption of nutrients. Evergreens, e.g., *Citrus* spp. and olive *(Olea europaea)* maintain their leaves for more than a year. Tree fruit species with deciduous habits retain leaves during the growing season but for less than 1 year, and edaphic variables (temperature, day length, drought, etc.) influence greatly the longevity of the canopy of trees cultivated in different climatic zones. Thus, in the Mediterranean climate of California, U.S., apple cultivars may retain their leaves for 8 months or longer while at the geographic limits of the north temperature zone (e.g., Minnesota, U.S.), the same cultivar may retain its foliage for less than 4 months.

A longer period of shoot extension and persistence of the leaf canopy in autumn is typical in young, i.e., vigorously growing trees and trees cultured under high levels of fertility and water availability.[16] On the other hand, the size of each leaf and total canopy leaf area will be affected adversely by water stress and/or nutrient deficiencies.[13,16,18,46] Not withstanding stress, the leaf mass of peach trees was reported to increase in direct proportion to tree size.[14]

Our frequent referral to the leaf canopy has been in the context of total absorptive surface and contains no implications for leaf homogeneity within the canopy. Thus, the development and properties of the cuticle which represents a principal barrier to the uptake of foliage-applied chemicals may vary greatly with species, leaf age, and environmental factors (light, temperature, humidity, etc.).[48]

Variability in light intensity and spectral distribution within the populations of leaves which comprise the canopy of fruit trees raises questions about the uniformity of the foliage within the canopy.[26,62] The distribution of light within the canopy can be influenced by a wide range of cultural practices — such as tree spacing, tree training and pruning, and choice of rootstock.[49] In general, leaf senescence is delayed in exposed vs. shaded portions of the canopy. Leaf nitrogen content and leaf photosynthetic capacity are not distributed uniformly over the canopies of fruit trees. Rather, this distribution is related closely to light exposure (T. M. DeJong, manuscript in preparation).

Leaves deficient in essential nutrients also reportedly have reduced capacities for photosynthesis,[107] but the effects of nutrient deficiencies on ion uptake following foliage application of nutrients have received limited attention. In olive, nitrogen deficiency reduced the redistribution of urea nitrogen from the leaves treated but did not affect nitrogen uptake.[57]

Leaf homogeneity may also vary as a function of proximity to developing fruit. We have obtained data in olive that redistribution of foliage-applied nitrogen from treated leaves is greatly influenced by the demand of metabolic sinks.[57]

At any time, depending upon species, the leaf canopy may represent a range of physiological leaf ages from immature, partly expanded leaves at the shoot tip to mature, fully expanded midshoot and spur leaves and, by midsummer to senescing basal leaves. This heterogeneity among leaves within the canopy is undoubtedly reflected by the variation among leaves in uptake and redistribution (export) of foliage-applied nutrients. Abaxial wax concentrations (micrograms per square centimeter) of the leaf surface appear to be diluted during leaf expansion and are also related inversely with temperatures during leaf development.[64]

Fisher[28] reported a greater response to foliage applications of urea made in June or later as compared with earlier applications. He interpreted this effect as a result of a greater leaf area available for urea absorption at the later dates. Also, recently expanded foliage is most sensitive to phytotoxicity,[28] and the export of foliage-applied nutrients from immature leaves is likely to be reduced[39] as early season applications result in greater incorporation into leaf protein from which there is slow turnover of individual nitrogen atoms during the summer and from which little nitrogen is translocated to the rest of the tree before leaf fall. In contrast, excess nitrogen applied via foliage in late season remains more mobile, and a high proportion of it moves to other tissues before leaf senescence.[42,57]

VII. FACTORS INFLUENCING NUTRIENT REDISTRIBUTION FROM LEAVES

A. Elemental Phloem Mobility

The efficacy of nutritional sprays would appear to be dependent not only on the absorption of foliage-applied nutrients through the cuticle and epidermal cells but also the transport of these nutrients to fruit and other plant parts.[10] Nutrients remobilized from senescing leaves are transported in the phloem[39] as presumably are foliage-applied nutrients redistributed from treated leaves. Thus, the phloem mobility of nutrient elements is an important aspect of the remedial effects of foliage applications of nutrient deficient plants, and factors influencing the extent of nutrient redistribution should be considered.

Materials mobile in the phloem are believed to move independently down concentration gradients from "source" (usually mature leaves) to "sinks" (immature and nonphotosynthesizing plant parts). Three general factors combine to determine the overall phloem mobility of a nutrient: (1) the ability of a nutrient to enter the phloem at the "source", (2) the ability of a nutrient to move within the phloem, and (3) the ability of a nutrient to move out of the phloem at the "sink". At least two aspects may be involved in the ability of a nutrient to enter the phloem at the "source" — the release of the nutrient from leaf cells and transport across membranes into the phloem.[39] The ability of a nutrient to move within the phloem may be a function of the ion or complex transported and/or solubility in the phloem sap.[32]

The documentation is overwhelming for remobilization of phloem-mobile nutrients (especially nitrogen, phosphorus, and potassium) from senescing leaves — as the net contents of mobile nutrients decline markedly (i.e., up to 50% or more) between leaf maturity and full senescence (Table 2). It is of interest that nutrients exhibiting the greatest phloem mobility represent the macroelements — essential nutrients required in greatest amounts by plants. Nutrient remobilization prior to leaf senescence appears to represent an important strategy for nutrient conservation and is dependent upon the mobility of these nutrients in the phloem.

Data which relate the phloem mobility of other essential nutrients such as copper, zinc, sulfur, manganese, and iron are relatively few and frequently contradictory.[39]

Since phosphate is the major anion in phloem sap, translocation of cations such as Ca^{++}, Ba^{++}, and Pb^{++}, which form phosphates of low solubility, are limited in the sieve tubes by the solubility product of the phosphate salt.[39] The ability of a nutrient to move away from unloading sites at the "sink" may be responsible for continued downward concentration gradients from the phloem to the sinks.[39]

As discussed by Swietlik and Faust,[108] the well-documented phloem immobility of boron and calcium is of great practical relevance with respect to the efficacy of foliage applications in fruit trees. Many physiological disorders of fruits are associated with low calcium levels,[7] but because of its phloem immobility, foliage-applied calcium is not redistributed from treated leaves to fruit. Rather, the calcium solution must be deposited on the fruit itself to

Table 2
LEAF NUTRIENT REMOBILIZATION: DIFFERENCES AMONG
NUTRIENTS BASED ON COMPARISON OF NUTRIENT
CONCENTRATIONS IN PRESENESCENT (GREEN) VS. SENESCING
LEAVES OF 'FUERTE' AVOCADO[a] AND 'GRAVENSTEIN' APPLE[b]

Species	Percentage change[c] in leaf concentration during leaf senescence							
	N	P	K	Ca	Mg	Fe	Mn	S
Avocado	−48.7	−54.9	−24.6	−1.7	−7.4	+22.3	+11.6	−21.7
Apple	−52.4	−27.4	−36.0	+18.4	NC	ND	ND	ND

[a]   Adapted from Cameron, S. H. et al.[11]
[b]   Oland, K.[87]
[c]   − = Net remobilization during leaf senescence; + = net accumulation during leaf senescence;
       NC = no change in leaf concentration detected during leaf senescence, ND = no data presented.

be effective in decreasing the occurrence of physiological disorders. Swietlik and Faust[108] calculated that the contribution of calcium sprayed on the apple fruit surface is about 15% of the total calcium needed in the fruit to prevent the development of physiological disorders.

B. Leaf Ontogeny

The influence of leaf ontogeny on phloem mobility of nutrients must be superimposed over basic differences in phloem mobility of essential nutrients and the nutrient demand of metabolic sinks. Three principal stages in the life of a leaf have been described.[39] They are

1.   Adolescence — a period of growth.
2.   Maturity — a period of constant weight after growth cessation.
3.   Leaf aging and senescence — a period of decline in leaf weight, internal disorganization, and nutrient salvage[13] prior to leaf abscission.

Phloem mobility and, therefore, the rate of nutrient transport from the treated leaf following foliage application is not constant during the life of a leaf. Thus, as it has long been recognized that the transport of organic assimilates varies greatly during leaf ontogeny, redistribution of nutrients from leaves may also vary with leaf ontogeny. During very early growth, the grape leaf depended entirely on the import of assimilates from more mature leaves. Assimilate export began when the leaf reached 30% of its final area, but the import of assimates continued until the leaf reached 50% of its final area.[59] In the soybean leaf the rate of assimilate export continued to increase until the leaf reached full expansion.[39]

In contrast to the pattern of accumulation and redistribution of phloem-mobile nutrients by leaves, phloem-immobile nutrients continue to accumulate in leaves throughout the life of the leaf. This continuing accumulation of nutrients in leaves results from a freely continuing import of nutrient from the xylem into leaf cells combined with a poor ability of the nutrient to enter the phloem.

C. Cropping

Net concentrations of phloem-mobile macroelements in leaves, i.e., nitrogen, phosphorus, and potassium, are typically, although not invariably, reduced in cropping as compared with noncropping situations (Table 3). We believe the available data indicate that phloem-mobile nutrients follow a circuitous path through the leaves and are preferentially mobilized from the leaves by the fruit via the phloem rather than proceeding directly to fruit in the transpiration stream.[69,119]

Table 3
INFLUENCE OF CROPPING ON LEAF NUTRIENT
COMPOSITION AT HARVEST

Species	Net changes in leaf nutrients associated with increased crop loads[a]								Ref.
	N	P	K	Ca	Mg	Mn	Fe	Zn	
Apple	+	CD	−	+	+	+	NC	ND	8
Citrus	−	−	−	+	+	ND	ND	ND	105
Pecan	−	NC	−	+	+	+	+	+	23
Pistachio	−	−	+[b]	NC	NC	NC	NC	NC	112
Peach	+	NC	−	+	+	+	NC	NC	76
Avocado	−	−	−	ND	ND	ND	ND	ND	24
Olive	−	NC	−	+	+	ND	ND	ND	27, 56

[a] Changes expressed as a percentage of leaf dry weight; + = increase; − = decrease; ND = no data presented; NC = no significant change; CD = conflicting data.

[b] Data based on cropping vs. noncropping limbs on the same trees.

The pronounced influence of developing fruit of European plum[12] *(Prunus domestica)* and pecan[106] *(Carya illinoensis)* on the redistribution of macronutrients is evidenced by the reduction in leaf potassium concentration which precedes the development of leaf scorch and dieback of branches which are characteristic symptoms of potassium deficiency in these species. These symptoms are almost always limited to heavily fruiting shoots. Concentration of potassium in fruit of European plum is almost independent of potassium concentration in the rest of the tree. Fruit samples from trees with leaf potassium ranging from 0.5% (very deficient) to nearly 4% potassium (very high) all have approximately 1% potassium in the dry fruit flesh. Thus, the amount of potassium removed from the orchard in harvested fruit depends only on crop size.[12]

Seed maturation in a number of agriculturally important annual plants is accompanied by a partial net retranslocation of mineral nutrients from the leaves to the seeds.[80] Sinclair and deWit[103] suggested that the nitrogen requirements of developing seeds of pulse and legume crops exceeds the supply capacity of the roots and that the nitrogen deficit triggers catabolism of nitrogen-rich leaf proteins and a net transfer of the resulting soluble nitrogen to the seeds. Decreases in leaf nitrogen and/or leaf chlorosis (e.g., in pistachio, Weinbaum unpublished) have also been associated with the proximity of developing fruit in tree fruit species (Table 3). Data obtained using ^{15}N-depleted $(NH_4)_2SO_4$ indicated that nitrogen flux through leaves to fruit and seed of almond *(Prunus dulcis)* may also occur with no net loss of nitrogen by the foliage.[119] These data appear consistent with the hypothesis that mature leaves remove organic nitrogenous constituents from xylem and transfer some of this nitrogen via the phloem to centers of growth or nitrogen accumulation such as seeds.[69] Redistribution of potassium to fruit from leaves via the phloem also appears likely to occur in crops in which fruit ripening takes place rapidly (e.g., European plum and pecan). In these cases, nutrient supply does not keep up with sink demand. If potassium is low in the soil solution, a temporary dip can be detected in grape leaf potassium at the veraison state (I. Klein, personal communication).

If correct, this concept is consistent with the idea that nutrient redistribution from leaves represents a normal channel of supply to fruit, at least for certain phloem-mobile nutrients. Thus, significant transport of phloem-mobile nutrients from leaves to fruit occurs irrespective of whether the nutrient originates in the soil or is contributed to the plant as a result of foliage application. As a result of the use of nitrogen isotopes, the concept appears well documented with respect to nitrogen,[69,90] but experimentation to date has not been adequate to resolve the phenomenon with respect to other phloem-mobile nutrients.

A positive correlation has been reported between the presence of developing fruit and transpiration rate (i.e., water consumption).[55] Greater accumulation of phloem-immobile nutrients (e.g., calcium) in leaves of cropping as compared with noncropping trees of the same cultivar may be related to these higher transpiration rates and resultant greater movement of nutrients to leaves in the transpiration stream (Table 3).

D. Nutritional Status of the Plant

Rapid uptake of foliage-applied nutrients and a temporary alleviation of boron, copper, calcium, iron, magnesium, manganese, potassium, nitrogen, and zinc deficiencies have long been appreciated in fruit trees.[108] The comparative utilization of foliage-applied nutrients by nutrient sufficient and deficient plants has received much less attention.[108] Cook and Boynton[19] reported a greater uptake rate of foliage-applied urea by apple leaves of higher nitrogen status but did not assess possible effects of nitrogen following uptake. We recently reported comparable absorption (milligrams nitrogen per plant) of foliage-applied urea by nitrogen-sufficient and nitrogen-deficient 'Manzanillo' olive plants, but our data indicated 17% greater retention of urea nitrogen by leaves of nitrogen-deficient plants than plants with adequate nitrogen status.[59]

The copper content of the oldest leaves in copper-deficient wheat plants declined much more slowly than that of copper-sufficient plants.[40] Hill et al.[40] suggested that deficiency of copper impaired the growth of new leaves and roots, thus removing the major sink for carbon and nitrogen translocated out of old leaves.

Clearly the reuse of copper deposited initially in old leaves is more critical for the growth of copper-deficient plants than of copper-sufficient plants. It is paradoxical, therefore, that severe copper deficiency delays the senescence of old leaves and prevents retranslocation of their copper. Perhaps the first priority in demand for nutrients is to maintain the functional integrity of the leaf. This would appear to have pronounced relevance to the alleviation of nutrient deficiency by foliage application of nutrients. The relationship between plant nutrient status and the redistribution of foliage-applied nutrients has received little attention. Attempts to supply the entire nitrogen regime of small apple trees by foliage application resulted in curiously abnormal plants.[29] Tree growth was stunted despite development of large leaves in which nitrogen levels appeared normal. The essentiality of soil nitrogen availability for normal root function, which includes hormone biosynthesis and transport, has received limited study. In grape, evidence of boron transport out of mature leaves was seen only when leaf boron concentrations were above a certain level.[100] Also, export of foliage-applied nitrogen from mature plum leaves was apparently enhanced when uptake of foliage-applied NO_3^- was accentuated using surfactants.[117]

Using ^{15}N-enriched urea, we have demonstrated the preferential export of foliage-applied nitrogen as compared with nonlabeled bulk leaf nitrogen.[57] This greater mobility of nitrogen recently applied to foliage as compared with bulk leaf nitrogen may represent deposition in different nitrogen pools. Ribulose-1,5 bisphosphate carboxylase (RuBP Case), the carboxylating enzyme, constitutes 40 to 60% of total soluble leaf protein in many species and is known to turn over slowly.[47]

VIII. FACILITATION OF UPTAKE AND TRANSPORT OF FOLIAGE-APPLIED CHEMICALS

Although plant growth regulators (PGR) are known to have dramatic and diverse effects on plant growth and metabolism, the effects of PGRs on the absorption and redistribution of foliage-applied nutrients have received little attention.[52] Halevy and Wittwer[36] reported that application of 10^{-5} *M* gibberellic acid (GA$_3$) to the root increased absorption of ^{86}Rb by the foliage but did not affect total translocation from the treated leaf. NAA (1-naphthalene acetic acid) applied at 10^{-6} M greatly enhanced uptake and transport of foliage-applied ^{86}Rb to the root. Their results suggested that absorption and the subsequent transport of foliage-applied rubidium were independent processes, and that the mobilization of rubidium by the various plant organs was not always determined by their respective growth rates.

Richards[93] reported that foliage application of 6-benzylaminopurine stimulated potassium uptake but depressed calcium uptake by roots of peach seedlings. The physiological basis of this phenomenon is beyond the scope of the present discussion, but there can be little doubt that there is a close relationship between shoot and root metabolism, the balance between them being regulated by the so-called plant hormones.[114] It has been appreciated for some that that substances like kinetin may enchance the mobilization of organic and inorganic substances to sites of application.[53,78] Kannan[53] anticipates an increased role of PGRs in the uptake and redistribution of foliage-applied nutrients. He believes this new technology has particular relevance to elements like iron, manganese, zinc, calcium, and boron which are not freely mobile within the plant.

As cuticular penetration represents a primary and major barrier to absorption of foliage-applied nutrients,[63] various attempts have been made to increase the permeability of the cuticle to foliage-applied chemicals. Cuticular penetration is a passive process and not energy-mediated, as the cuticle is devoid of living cells.[52] Enhancement of cuticular penetration of ions in the presence of urea has been attributed to "facilitated diffusion".[3,52] DMSO (dimethylsulfoxide), a commonly used solvent, used at 0.5 or 1% increased absorption of [59]$FeSO_4$ by corn plants.[15]

Stomata potentially play a twofold role in foliar penetration of chemicals. The guard and accessory cells per se have been shown to be preferential sites for foliar absorption[30] of organic molecules perhaps because they are uncutinized, at least in some species.[1] Uptake of 2,4-D was proportional to the density of stomata.[98] Stomata have also been implicated in the uptake of other herbicides such as picloram,[97] bentazone[74] and triclopyr.[54] Uptake of herbicides has been shown to increase when the stomata are open, probably because the pore provides access for solutions to areas where the cuticle is thinner and where less wax is deposited. Stomatal closure in dry seasons has been suggested to explain the reduced uptake of herbicides during those periods as it has been proposed that uptake through the wax-free stomatal antechamber might be the principal route by which bentazone enters the mature leaf.[74] Preliminary evidence suggests that indol-3-ylacetic acid (IAA) can stimulate stomatal opening at times when it would otherwise be inhibited by other factors, thus providing the possibility of regulating stomatal aperture in the field to facilitate the uptake of applied chemicals.[74] A number of additives (IAA, NAA, TIBA, DNP, and urea) to $ZnSO_4$ sprays each enhanced zinc uptake by 50 to 100%,[3,43] but it is unclear whether the effect is mediated by increased stomatal aperture, increased membrane permeability, or other parameters.

There are apparently no data indicating that absorption by stomatal guard cells influence significantly the uptake of foliage-applied nutrients.

The second role that stomata may play in the absorption of foliage-applied chemicals under certain conditions is the mass movement of the spray solution through the stomatal pore into the substomatal chamber.[33] The substomatal cavity has a relatively large surface area, and solutions which enter it effectively by-pass the thick external cuticle and need only traverse the relatively thin cuticle lining the cavity[31] to reach the mesophyll. Surface tensions below 30 dynes per centimeter permit penetration of aqueous solutions through stomata into substomatal cavities.[99] Some fluorocarbon and silicone polymer-based surfactants reduce surface tensions of aqueous solutions below 30 dynes per centimeter without phytotoxicity,[81] and a silicone-based surfactant (L77) greatly increased initial absorption of phosphate ($\times 10$) and iron ($\times 3$) by bean leaves from aqueous solutions of their salts.[82] L77 also doubled initial uptake of NO_3^- by leaves of European plum.[117] The newer organosilicone-based sufactants have also given encouraging results in field trials with herbicide sprays,[51] but did not enhance urea penetration in olive.[58] As a small, nonpolar molecule, urea penetration of the leaf cuticle may be sufficiently rapid[124] that any advantage derived from stomatal penetration is precluded.

ACKNOWLEDGMENTS

The author acknowledges useful discussions with Dr. R. M. Carlson and critical reading of the chapter by Dr. I Klein.

REFERENCES

1. **Appleby, R. F. and Davies, W. J.,** A possible evaporation site in the guard cell wall and the influence of leaf structure on the humidity response by stomata of woody plants, *Oecologia,* 56, 30, 1983.
2. **Arnon, D. I. and Stout, P. R.,** The essentiality of certain elements in minute quantity for plants with special reference to copper, *Plant Physiol.,* 14, 371, 1939.
3. **Bar-Akiva, A. and Hewitt, E. J.,** The effects of triiodobenzoic acid and urea on the response of chlorotic lemon *(Citrus liminia)* trees to foliar application of iron compounds, *Plant Physiol.,* 34, 641, 1959.
4. **Barr, W. and Pellett, H.,** Effect of soil temperature on growth and development of some woody plants, *J. Am. Soc. Hortic. Sci.,* 97, 632, 1972.
5. **Batjer, L. P., Magness, J. R., and Regeimbal, L. O.,** The effect of root temperature on growth and nitrogen intake of apple trees, *Proc. Am. Soc. Hortic. Sci.,* 37, 11, 1940.
6. **Beutel, J., Uriu, K., and Lilleland, O.,** Leaf analysis for California deciduous fruits, in *Soil and Plant Tissue Testing in California, Bulletin 1879,* University of California, Div. of Agric. Sci., 1983, 15.
7. **Beyers, E.,** Control of bitter pit and other disorders of apples with calcium sprays, *Decid. Fruit Grower,* 13, 319, 1963.
8. **Bould, C.,** Leaf analysis of deciduous fruits, in *Nutrition of Fruit Crops,* Childers, N. F., Ed., Horticultural Publications, Rutgers — The State University, New Brunswick, N.J., 1966, 651.
9. **Boynton, D., Margolis, D., and Gross, C. R.,** Exploratory studies on nitrogen metabolism by McIntosh apple leaves sprayed with urea, *Proc. Am. Soc. Hortic. Sci.,* 62, 135, 1953.
10. **Bukovac, M. J. and Wittwer, S. H.,** Absorption and mobility of foliar applied nutrients, *Plant Physiol.,* 32, 428, 1957.
11. **Cameron, S. H., Mueller, R. T., and Wallace, A.,** 1952 Nutrient composition and seasonal losses of avocado trees, *Calif. Avocado Soc. Yearb.,* 37, 201, 1952.
12. **Carlson, R. M. and Uriu, K.,** Potassium in prune production, in *Prune Orchard Management,* Special Publ. 3269, Division of Agric. Sci., University of California, 1981, 98.
13. **Chabot, B. F. and Hicks, D. J.,** The ecology of leaf life spans, *Annu. Rev. Ecol. Syst.,* 13, 229, 1982.
14. **Chalmers, D. J. and van den Ende, B.,** Productivity of peach trees: factors affecting dry-weight distribution during tree growth, *Ann. Bot.,* 39, 423, 1975.
15. **Chamel, A.,** Penetration et migration du ^{59}Fe applique sur les feuilles de Maïs; effet du dimethyl sulfoxyde, *Physiol. Plant.,* 26, 170, 1972.
16. **Chandler, W. H.,** *Deciduous Orchards,* 3rd ed., Lea & Febiger, Philadelphia, 1957.
17. **Chandler, W. H. and Brown, D. S.,** Deciduous orchards in California winters, *Calif. Agric. Exp. Stn. Circ.,* 179, 1, 1951.
18. **Childers, N. F., Ed.,** *Nutrition of Fruit Crops,* Horticultural Publications, Rutgers — The State University, New Brunswick, N.J., 1966.
19. **Cook, J. A. and Boynton, D.,** Some factors affecting the absorption of urea by McIntosh apple leaves, *Proc. Am. Soc. Hortic. Sci.,* 59, 82, 1952.
20. **Cooke, G. W.,** The energy costs of nitrogen fertilizers used in Britain, the returns received and some savings that are possible, *J. Sci. Food Agric.,* 26, 1065, 1975.
21. **Dalbro, S.,** Leaching of apple foliage by rain, *Rep. 14th Int. Hortic. Congr.,* 1, 770, 1955.
22. **Delap, A. V.,** The response of young apple trees of differing nitrogen status to a urea spray in autumn, *Annu. Rep. E. Malling Res. Stn.,* 1966, 139, 1967.
23. **Diver, S. G., Smith, M. W., and McNew, R. W.,** Influence of fruit development on seasonal elemental concentrations and distribution in fruit and leaves of pecan, *Commun. Soil Sci. Plant. Anal.,* 15, 619, 1984.

24. **Embleton, T. W. and Jones, W. W.,** *Nutrition of Fruit Crops,* Childers, N. F., Ed., Horticultural Publications, Rutgers — The State University, New Brunswick, N.J., 1966, 51.

25. **Embleton, T. W. and Jones, W. W.,** Foliar-applied nitrogen for citrus fertilization, *J. Environ. Qual.,* 3, 388, 1974.

26. **Erez, A. and Kadman-Zahavi, A.,** Growth of peach plants under different filtered sunlight conditions, *Physiol. Plant.,* 26, 210, 1972.

27. **Fahmy, I. and Nasrallah, S.,** Changes in macro-nutrient elements of Souri olive leaves in alternate bearing years, *Proc. Am. Soc. Hortic. Sci.,* 74, 372, 1959.

28. **Fisher, E. G.,** The principles underlying foliage applications of urea for nitrogen fertilization of the McIntosh apple, *Proc. Am. Soc. Hortic. Sci.,* 59, 91, 1952.

29. **Forshey, C. G.,** A comparison of soil nitrogen fertilization and urea sprays as sources of nitrogen for apple trees in sand culture, *Proc. Am. Soc. Hortic. Sci.,* 83, 32, 1963.

30. **Franke, W.,** Role of guard cells in foliar absorption, *Nature,* 202, 1236, 1964.

31. **Franke, W.,** Mechanisms of foliar penetration of solutions, *Annu. Rev. Plant Physiol.,* 18, 281, 1967.

32. **Van Goor, G. J. and Wiersma, D.,** Redistribution of potassium, calcium, magnesium, and manganese in the plant, *Physiol. Plant.,* 31, 163, 1974.

33. **Greene, D. W. and Bukovac, M. J.,** Stomatal penetration: effect of surfactants and role in foliar absorption, *Am. J. Bot.,* 61, 100, 1974.

34. **Gur, A., Bravdo, B., and Mizrahi, Y.,** Physiological responses of apple trees to supraoptimal root temperature, *Physiol. Plant.,* 27, 130, 1972.

35. **Gur, A., Hepner, J., and Shulman, Y.,** The influence of root temperature on apple trees. IV. The effect on the mineral nutrition of the tree, *J. Hortic. Sci.,* 54, 313, 1979.

36. **Halevy, A. H. and Wittwer, S. H.,** Foliar uptake and translocation of rubidium in bean plants as affected by root absorbed growth regulators, *Planta,* 67, 375, 1965.

37. **Hanson, E. J., Chaplin, M. H., and Breen, P. J.,** Movement of foliar applied boron out of leaves and accumulation in flower buds and flower parts of 'Italian' prunes, *Hortic. Sci.,* 20, 747, 1985.

38. **Haynes, R. J. and Goh, K. M.,** Review of physiological pathways of foliar absorption, *Sci. Hortic.,* 7, 291, 1977.

39. **Hill, J.,** The remobilization of nutrients from leaves, *J. Plant Nutr.,* 2, 407, 1980.

40. **Hill, J., Robson, A. D., and Loneragan, J. F.,** The effects of copper supply and shading on retranslocation of copper from mature wheat leaves, *Ann. Bot.,* 43, 449, 1979.

41. **Hill-Cottingham, D. G.,** Effect of time of application of fertilizer nitrogen on the growth, flowering and fruiting of maiden apple trees grown in sand culture, *J. Hortic. Sci.,* 38, 242, 1963.

42. **Hill-Cottingham, D. G. and Lloyd-Jones, C. P.,** Nitrogen-15 in apple nutrition investigations, *J. Sci. Food Agric.,* 26, 165, 1975.

43. **Hoffman, M. and Samish, R. M.,** The control of zinc deficiency in apple, *Israel J. Agric. Res.,* 16, 105, 1966.

44. **Holloway, P. J.,** Surface factors affecting the wetting of leaves, *Pestic. Sci.,* 1, 156, 1970.

45. **Horesh, I. and Levy, Y.,** Response of iron-deficient citrus trees to foliar iron sprays with a low surface-tension surfactant, *Sci. Hortic.,* 15, 227, 1981.

46. **Hsiao, T. C.,** Plant responses to water stress, *Annu. Rev. Plant Physiol.,* 24, 519, 1973.

47. **Huffaker, R. C.,** Biochemistry and physiology of leaf proteins, in *Encyclopedia of Plant Physiology,* Vol. 14A, Boulter, I. D. and Parthier, B., Eds., Springer-Verlag, Heidelberg, 1982, 370.

48. **Hull, H. M., Morton, H. L., and Wharrie, J. R.,** Environmental influences on cuticle development and resultant foliar penetration, *Bot. Rev.,* 41, 421, 1975.

49. **Jackson, J. E.,** Light interception and utilization by orchard systems, *Hortic. Rev.,* 2, 208, 1980.

50. **Jackson, M. B.,** Approaches to relieving aeration stress in waterlogged plants, *Pestic. Sci.,* 14, 25, 1983.

51. **Jansen, L. L.,** Enhancement of herbicides by silicone surfactants, *Weed Sci.,* 21, 130, 1973.

52. **Kannan, S.,** Mechanisms of foliar uptake of plant nutrients — accomplishments and prospects, *J. Plant Nutr.,* 2, 717, 1980.

53. **Kessler, B. and Moscicki, Z. W.,** Effects of triiodobenzoic acid and maleic hydrazide upon the transport of foliar applied calcium and iron, *Plant Physiol.,* 33, 70, 1958.

54. **King, M. G. and Radosevich, S. R.,** Tanoak *(Lithocarpus densiflorus)* leaf surface characteristics and absorption of Triclopyr, *Weed Sci.,* 27, 599, 1979.

55. **Klein, I.,** Drip irrigation based on soil matric potential conserves water in peach and grape, *Hortic. Sci.,* 18, 942, 1983.

56. **Klein, I. and Lavee, S.,** The effect of nitrogen and potassium fertilizers on olive production, in *Fertilizer Use and Production of Carbohydrates and Lipids,* 13th Colloq. Potash Inst., York, England, 1977, 295.

57. **Klein, I. and Weinbaum, S. A.,** Foliar application of urea to olive: translocation of urea nitrogen as influenced by sink demand and nitrogen deficiency, *J. Am. Soc. Hortic. Sci.,* 109, 356, 1984.

58. **Klein, I. and Weinbaum, S. A.,** Foliar application of urea to almond and olive: leaf retention and kinetics of uptake, *J. Plant Nutr.,* 8, 117, 1985.

59. **Koblet, W.,** Wanderung von Assimilaten in Rebtrieben und Einfluss der Blattflache auf Ertrag und Qualitat der Trauben, *Wein-Wiss.,* 24, 277, 1969.
60. **Kramer, P. J.,** Root resistance as a cause of decreased water absorption by plants at low temperature, *Plant Physiol.,* 15, 63, 1940.
61. **Labanauskas, C. K., Stolzy, L. H., and Handy, M. F.,** Concentrations and total amounts of nutrients in citrus seedlings (*Citrus sinensis* 'Osbeck') and in soil as influenced by differential soil oxygen treatments, *Soil Sci. Soc. Am. Proc.,* 36, 454, 1972.
62. **Lakso, A. N.,** Correlations of fisheye photography to canopy structure, light climate, and biological responses to light in apple tree, *J. Am. Soc. Hortic. Sci.,* 105, 43, 1980.
63. **Leece, D. R.,** Composition and ultrastructure of leaf cuticles from fruit trees, relative to differential foliar absorption, *Aust. J. Plant Physiol.,* 3, 833, 1976.
64. **Leece, D. R.,** Foliar absorption in *Prunus domestica* L. I. Nature and development of the surface wax barrier, *Aust. J. Plant Physiol.,* 5, 749, 1978.
65. **Leece, D. R. and Dirou, J. F.,** Organosilicone and alginate adjuvants evaluated in urea sprays foliar-applied to prune trees, *Commun. Soil Sci. Plant Anal.,* 8, 169, 1977.
66. **Leece, D. R. and Dirou, J. F.,** Comparison of urea foliar sprays containing hydrocarbon or silicone surfactants with soil-applied nitrogen in maintaining the leaf nitrogen concentration of prune trees, *J. Am. Soc. Hortic. Sci.,* 104, 644, 1979.
67. **Leece, D. R. and Kenworthy, A. L.,** Effect of potassium nitrate foliar sprays on leaf nitrogen concentrations and growth of peach trees, *Hortic. Science,* 6, 171, 1971.
68. **Leece, D. R. and Kenworthy, A. L.,** Influence of epicuticular waxes on foliar absorption of nitrate ions by apricot leaf disks, *Aust. J. Biol. Sci.,* 25, 641, 1972.
69. **Lewis, O. A. M. and Pate, J. S.,** The significance of transpirationally derived nitrogen in protein synthesis in fruiting plants of pea (*Pisum sativim* L.), *J. Exp. Bot.,* 24, 596, 1973.
70. **Lilleland, O.,** Potassium deficiency of fruit trees in California, *Proc. XV Int. Hortic. Congr.,* Nice, 168, 1961.
71. **Little, R. C., Charlesworth, R. R., and Roach, F. A.,** Postharvest urea spraying of apples, *Exp. Hortic.,* 15, 27, 1966.
72. **Looney, N. E.,** Growth regulator usage in apple and pear production, *Plant Growth Regulating Chemicals,* Vol. I., Nickell, L. G., Ed., CRC Press, Boca Raton, Fla., 1983, 1.
73. **Mann, C. E. T. and Wallace, T.,** The effect of leaching with cold water on the foliage of the apple, *J. Pomol. Hortic. Sci.,* 4, 146, 1925.
74. **Mansfield, T. A., Pemadasa, M. A., and Snaith, P. J.,** New possibilities for controlling foliar absorption via stomata, *Pestic. Sci.,* 14, 294, 1983.
75. **Miller, R. J. and Smith, R. B.,** Nitrogen balance in the southern San Joaquin Valley, *J. Environ. Qual.,* 5, 274, 1976.
76. **McClung, A. C. and Lott, W. L.,** Mineral nutrient composition of peach leaves as affected by leaf age and position and the presence of a fruit crop, *Proc. Am. Soc. Hortic. Sci.,* 67, 113, 1956.
77. **Moreno, J. and Garcia-Martinez, J. L.,** Nitrogen accumulation and mobilization in *Citrus* leaves throughout the annual cycle, *Physiol. Plant.,* 61, 429, 1984.
78. **Muller, K. and Leopold, A. C.,** The mechanisms of kinetin induced transport in corn leaves, *Planta,* 68, 186, 1966.
79. **Nielsen, G. H. and Hogue, E. J.,** Foliar application of chelated and mineral zinc sulphate to Zn-deficient 'McIntosh' seedlings, *Hortic. Sciences,* 18, 915, 1983.
80. **Neumann, P. M.,** Late season foliar fertilization with macronutrients — is there a theoretical basis for increased seed yields?, *J. Plant Nutr.,* 5, 1209, 1982.
81. **Neumann, P. M. and Prinz, R.,** Evaluation of surfactants for use in the spray treatment of iron chlorosis in citrus trees, *J. Sci. Food Agric.,* 25, 221, 1974.
82. **Neumann, P. M. and Prinz, R.,** The effect of organosilicone surfactants in foliar nutrient sprays on increased absorpiton of phosphate and iron salts through stomatal infiltration, *Israel J. Agric. Res.,* 23, 123, 1974.
83. **Norton, R. A. and Childers, N. F.,** Experiments with urea sprays on the peach, *Proc. Am. Soc. Hortic. Sci.,* 63, 23, 1954.
84. **Oland, K.,** Nitrogenous reserves of apple trees, *Physiol. Plant.,* 13, 594, 1959.
85. **Oland, K.,** Nitrogen feeding of apple trees by post-harvest urea sprays, *Nature,* 185, 856, 1960.
86. **Oland, K.,** Responses of cropping apple trees to post-harvest urea sprays, *Nature, (London)* 198, 1282, 1963.
87. **Oland, K.,** Changes in the content of dry matter and major nutrient elements of apple foliage during senescence and abscission, *Physiol. Plant.,* 16, 682, 1963.
88. **Page, A. L., Martin, J. P., and Ganje, T. J.,** Foliar absorption and translocation of potassium by *Citrus,* *Proc. Am. Soc. Hortic. Sci.,* 82, 165, 1963.

89. **Paris, I.**, The biological importance of titanium, *J. Plant Nutr.*, 6, 3, 1983.
90. **Pate, J. S., Layzell, D. B., and McNeil, D. L.**, Modeling the transport and utilization of carbon and nitrogen in a nodulated legume, *Plant Physiol.*, 63, 730, 1979.
91. **Price, C. E. and Anderson, N. H.**, Uptake of chemicals from foliar deposits: effects of plant species and molecular structure, *Pestic. Sci.*, 16, 369, 1985.
92. **Preistly, C. A.**, The annual turnover of resources in young olive trees, *J. Hortic. Sci.*, 52, 105, 1977.
93. **Richards, D.**, Root-shoot interactions: functional equilibria for nutrient uptake in peach (*Prunus persica* L., Batch), *Ann. Bot.*, 42, 1039, 1978.
94. **Rowe, R. N. and Beardsell, D. V.**, Waterlogging of fruit trees, *Hortic. Abstr.*, 43, 533, 1973.
95. **Ryugo, K.**, The Adequacy of Photosynthesis to Supply the Demand of the Maturing Peach Fruit, Doctoral dissertation, University of California, Davis, 1954.
96. **Salardini, A. A.**, Yield response of tea (*Camelia sinesis*) to fertilizer and irrigation in two locations of Iran, *Plant Soil*, 50, 113, 1978.
97. **Sands, R. and Bachelard, E. P.**, Uptake of picloram by Eucalypt leaf discs. II. Role of stomata, *New Phytol.*, 72, 87, 1973.
98. **Sargent, J. A. and Blackman, G. E.**, Studies on foliar penetration. I. Factors controlling the entry of 2,4-dichloro phenoxyacetic acid, *J. Exp. Bot.*, 13, 348, 1962.
99. **Schönherr, J. and Bukovac, M. J.**, Penetration of stomata by liquids. Dependence on surface tension, wettability, and stomatal morphology, *Plant Physiol.*, 49, 813, 1972.
100. **Scott, L. E. and Schrader, A. L.**, Effect of alternating conditions of boron nutrition upon growth and boron content of grape vines in sand culture, *Plant Physiol.*, 22, 526, 1947.
101. **Shim, K. K., Titus, J. S., and Splittstoesser, W. E.**, The utilization of post-harvest urea sprays by senescing apple leaves, *J. Am. Soc. Hortic. Sci.*, 97, 592, 1972.
102. **da Silva, P. R. F. and Stutte, C. A.**, Nitrogen volatilization from rice leaves. I. Effects of genotype and air temperature, *Crop. Sci.*, 21, 596, 1981.
103. **Sinclair, T. R. and deWit, C. T.**, Photosynthate and nitrogen requirements for seed production by various crops, *Science*, 189, 565, 1975.
104. **Smith, A.**, Seasonal subsoil temperature variations, *J. Agric. Res.*, 44, 421, 1932.
105. **Smith, P.**, Leaf analysis of citrus. in *Nutrition of Fruit Crops*, Childers, N. F., Ed., Horticultural Publications, Rutgers — The State University, New Burnswick, N.J., 1966, 208.
106. **Sparks, D.**, Effects of fruiting on scorch, premature defoliation, and nutrient status of 'Chickasaw' pecan leaves, *J. Am. Soc. Hortic. Sci.*, 102, 669, 1977.
107. **Spiller, S. and Terry, N.**, Limiting factors in photosynthesis. II. Iron stress diminishes photochemical capacity by reducing the number of photosynthetic units, *Plant Physiol.*, 65, 121, 1980.
108. **Swietlik, D. and Faust, M.**, Foliar nutrition of fruit crops, *Hortic. Rev.*, 6, 287, 1984.
109. **Titus, J. S. and Kang, S. M.**, Nitrogen metabolism, translocation, and recycling in apple trees, *Hortic. Rev.*, 4, 204, 1982.
110. **Tromp, J.**, The effect of root temperature on the absorption and distribution of K, Ca, and Mg in three rootstock clones of apple budded with Cox's Orange Pippin, *Gartenbauwissenschaft*, 43, 49, 1978.
111. **Tromp, J.**, Mineral absorption and distribution in young apple tree under various environmental conditions, in *Mineral Nutrition of Fruit Trees*, Atkinson, D., Ed., Butterworths, Boston, 1980, 173.
112. **Uriu, K. and Crane, J. C.**, Mineral element changes in pistachio leaves, *J. Am. Soc. Hortic. Sci.*, 102, 155, 1977.
113. **Uriu, K., Carlson, R. M., Henderson, D. W., Schulbach, H., and Aldrich, T. M.**, Potassium fertilization of prune trees under drip irrigation, *J. Am. Soc. Hortic. Sci.*, 105, 508, 1980.
114. **Vaadia, Y. and Itai, C.**, Interrelationships of growth with reference to the distribution of growth substances, in *Root Growth*, Whittington, W. J., Ed., Butterworths, London, 1969, 65.
115. **Weiland, R. T. and Stutte, C. A.**, Concomitant determination of foliar nitrogen loss, net carbon dioxide uptake and transpiration, *Plant Physiol.*, 65, 403, 1980.
116. **Weinbaum, S. A.**, Feasibility of satisfying total nitrogen requirement of non-bearing prune tree with foliar nitrate, *Hortic. Sci.*, 13, 52, 1978.
117. **Weinbaum, S. A. and Neumann, P. M.**, Uptake and metabolism of ^{15}N-labeled potassium nitrate by French prune (*Prunus domestica* L.) leaves and the effects of two surfactants, *J. Am. Soc. Hortic. Sci.*, 102, 601, 1977.
118. **Weinbaum, S. A., Klein, I., Broadbent, F. E., Micke, W. C., and Muraoka, T. T.**, Use of isotopic nitrogen to demonstrate dependence of mature almond trees on annual uptake of soil nitrogen, *J. Plant Nutr.*, 7, 975, 1984.
119. **Weinbaum, S. A. and Muraoka, T. T.**, Nitrogen redistribution from almond foliage and pericarp to the almond embryo, *J. Am. Soc. Hortic. Sci.*, in press, 1986.
120. **Wiersum, L. K.**, Soil water content in relation to nutrient uptake by the plant. *Versl. Meded. Comm. Hydrol. Onderz.* TNO, 15, 74, 1969.

121. **Wiersum, L. K.,** The effect of soil physical conditions on roots and uptake, in *Mineral Nutriton of Fruit Trees,* Atkinson, D., Ed., Butterworths, Boston, 1980, 111.
122. **Wilson, B. F.,** *The Growing Tree,* University of Massachusetts Press, Amherst, 1970.
123. **Wilson, W. S.,** Nitrogen manuring of apple trees via root and leaf: a preliminary investigation, *Exp. Hortic.,* 15, 33, 1966.
124. **Wittwer, S. H., Bukovac, M. J. and Tukey, H. B.,** Advances in foliar feeding of plant nutrients, in *Fertilizer Technology and Usage,* McVikar, M. H., Bridger, G. L., and Nelson, L. B., Eds., *Soil Science Society of America,* Madison, Wis., 1963, 429.
125. **Yogaratnam, N. and Greenham, D. W. P.,** The application of foliar sprays containing nitrogen magnesium, zinc and boron to apple trees. I. Effects on fruit set and cropping, *J. Hortic. Sci.,* 57, 151, 1982.

Chapter 6

FOLIAR APPLICATIONS OF FERTILIZERS ON GRAIN CROPS

John J. Hanway

TABLE OF CONTENTS

I. INTRODUCTION

Plants are capable of absorbing nutrients through aerial plant parts, so foliar application can be an effective method of applying fertilizer nutrients to plants under some conditions.[1,2] However, the amount of fertilizer that can be applied per application is limited because severe leaf burn results when too much fertilizer salts are applied at any one time. Therefore, foliar fertilization has been used primarily for applications of minor elements that are needed only in small amounts.

Foliar applications of the micronutrients are made extensively for deciduous fruit trees, citrus, grape vineyards, and for other crops such as pineapples, sugar cane, beans, and tomatoes. Several of the micronutrients such as iron (Fe), zinc (Zn), copper (Cu), and manganese (Mn) are applied primarily as sulfates. Boron (B) is applied as boric acid or borox, and molybdenum (Mo) is applied as molybdate. The amounts of the nutrients needed and used are too small to cause serious leaf burn.

Foliar fertilization with the macronutrients is less extensively used. Applications of magnesium (Mg) and sulfur (S) may be made using magnesium sulfate or other Mg and S compounds. These have been used primarily for tree and vegetable crops. Foliar applications of nitrogen (N), phosphorus (P), and potassium (K) can be effective in colder regions or seasons where nutrient uptake by roots is limited by cold temperatures. But, even in warmer climates, e.g., Hawaii, foliar applications of N, P, and K are used regularly for pineapple and sugar cane. Because the amounts of these elements that can be applied at one time is limited, successive applications throughout the season are required.

Foliar application of the major fertilizer elements — N, P, K, and S — has not been practiced extensively for grain crops because of the limitations due to the amounts required and the amounts that can be applied per application. However, recent research indicates a potential for foliar fertilization with these nutrient elements during the seed-filling period if this practice can be made consistenlty effective.

II. BASIS FOR FOLIAR FERTILIZATION DURING SEED-FILLING

During the seed-filling period in grain crops, soluble carbohydrates and several nutrient elements are translocated from the leaves and other vegetative plant parts to the developing seeds. This results in severe depletion of some of these elements in the leaves, and, as this nutrient depletion continues, photosynthesis in the leaves slows and stops. After that occurs, the only source of photosynthetic materials to the seeds is that which has accumulated in the plants, so the rate of seed-weight increase decreases markedly. Furthermore, there is little or no translocation of carbohydrates and nutrients from other plant parts to the plant roots during this period so growth and extension of the roots stops and N-fixation in the nodules on leguminous plants ceases.[3,4] So, the rate of nutrient uptake by the roots and translocation from the roots gradually decrease during seed-filling.

Soluble carbohydrates in the plants result from photosynthesis which occurs primarily in the leaves. These carbohydrates are translocated to various plant parts and are used in the growth of the plants, but in most plant parts some of the carbohydrates remain in soluble forms that can be retranslocated. As the seeds develop, these carbohydrates are translocated primarily to the developing seeds. If photosynthesis stops during the seed-filling period, seed-filling continues but the rate of seed-filling is slower and the final seed yield is much reduced. When this happens the final seed size and/or the number of filled seeds is reduced.

Foliar fertilization during seed-filling can supply the nutrients that are being depleted from the leaves during seed-fillings.[5] However, recent studies indicate that the foliarly applied nutrients are rapidly translocated from the leaves and do not maintain adequate nutrient contents in the leaves.

Table 1
COMMON CONCENTRATIONS OF
DIFFERENT ELEMENTS IN THE SEEDS
AND LEAVES OF SOYBEANS AND CORN

Chemical element	Concentration[a]			
	Seeds		Leaves	
	Soybeans	Corn	Soybeans	Corn
N	6.5	1.6	4.7	3.0
P	0.6	0.35	0.35	0.35
S	0.3	0.12	0.3	0.2
K	1.8	0.4	1.8	2.0
Ca	0.3	0.01	1.7	0.6
Mg	0.35	0.10	0.6	0.5

[a] Percentage on a dry weight basis.

III. EXPERIMENTAL RESULTS — PROBLEMS AND SOLUTIONS

Although the processes occurring in different grain-crop plants during seed-filling are basically similar, the concentrations of several of the essential elements in the seeds of soybeans are much higher than those in most other grain crops. For example, as shown in Table 1, the concentrations of six major elements are much higher in soybean seeds than in corn seeds although the concentrations in the leaves of these two crop plants are not that different. Depletion of N, P, and K from the leaves during seed-filling is severe in soybeans. Partly because of this, there has been more research on foliar fertilization of soybeans during seed-filling than on corn or other crops.

IV. SOYBEANS

Yield increases were obtained in some field experiments in 1974 and 1975 by Garcia and Hanway[6] with foliar applications of NPKS solutions during the seed-filling period. This occurred with several soybean cultivars. Yield increases of as much as 1570 kg/ha over a check yield of 3540 kg/ha were obtained. The fertilizer nutrients were applied as urea, potassium polyphosphate, and potassium sulfate. The foliar spray applications were most effective when the spray solution contained N + P + K + S in a ratio of 10:1:3:0.5, similar to that in soybean seeds, and was applied at a rate to supply a total of 80 + 9 + 24 + 4 kg/ha of N + P + K + S in four sprayings during the seed-filling period (between plant developmental stages R5 and R7).[7] If any of the nutrient elements was omitted from the NPKS spray solution, little or no yield increase resulted from the foliar application. Applications of more than 20 + 2 + 6 + 1 in one spray application caused serious leaf burn. Yield increases resulted from increases in numbers of seeds, not increases in seed size.

A similar NPKS foliar application by Vasilas et al.[8] at Beltsville, Md. resulted in a very significant yield increase of 1048 kg/ha over a check yield of 3203 kg/ha in 1977, but no yield increase in 1976. Other investigators[9-13] obtained no yield increases from such foliar applications. El-Hout[14] obtained a very significant yield increase with one soybean cultivar but no yield increase with two other cultivars. These inconsistent results from NPKS foliar fertilization of soybeans during the seed-filling period indicate that the effectiveness of this practice is influenced markedly by one or more other factors.

A. Forms of Nitrogen, Phosphorus, Potassium, and Sulfur

Wittwer and Bukovac[2] provide an excellent review of studies that had been conducted on nutrient absorption through leaf surfaces. But relatively little research had been conducted on field grain crops during the seed-filling period.

Urea has generally been found to be the least toxic and most effective form of N for foliar application. Urea is nonpolar and undissociated, is absorbed more rapidly (but 10- to 20-fold) than cations, and is highly mobile in the plants. However, many inorganic N (ammonium and nitrate) salts have been used successfully for foliar applications.

Barel and Black[15,16] tested the suitability of 32 different P compounds for foliar applications on corn and soybeans. They found the condensed P sources (tri- and tetrapolyphosphates) to be more effective than orthophosphates and other inorganic P sources for foliar application.

Potassium from various inorganic salts exists as the K^+ ion in water solution. When applied to plant foliage it has been found to be rapidly absorbed and translocated.[2]

Sulfur has usually been foliarly applied as a sulfate. It has been found to be absorbed slower than urea but faster than P.[17]

B. Time of Application

Although it has been shown that absorption of nutrients by leaves is generally greatest during daylight,[18] because of temperature effects (slow drying and dew formation) spraying of fertilizer solutions in the evening has often been shown to give the best results.[19] Where the effects of foliar applications of NPKS solutions made between midmoring and midafternoon have been compared with similar applications made earlier or later in the day, much greater leaf burn has often resulted from the midday applications.[20] Therefore, it is generally recommended that foliar fertilizer applications be made before 0800 or after 1700 hr.

For foliar fertilization of soybeans with an NPKS solution to be effective in increasing yield it must be applied during the seed-filling period — between growth stages R5 and R7.[7] Applications later in this period (after R6) have been more effective than those made earlier in the period (near R5).[20] Because yield increases have usually been associated with increases in numbers of harvested seeds rather than increases in size of the harvested seeds, it is apparent that for foliar fertilization during seed-filling to be successful more seeds must be initiated and filled than would normally be the case.

C. Rates of Application

Since a primary objective of foliar fertilization during seed-filling is to provide nutrients to the developing seeds, the rate of application required is related to the amounts of nutrients in the fully developed seeds. Soybean seeds normally contain about 6.5% N. This is much higher than the concentrations of the other nutrient elements in soybean seeds and of the N concentrations in most other grains. So, the amounts of N to be applied as a foliar spray during seed-filling are high for soybeans and limit the rate of application more than the amounts of other elements. Leaf burn has been shown to occur with an NPKS spray when no leaf burn resulted from a similar PKS spray without the N.[6]

Garcia[21] emphasized that in his studies application rates of more than 20 kg of N per hecture at any one spraying resulted in serious leaf burn and should be avoided. Similar results have been reported by other investigators, but serious burn has occurred in some experiments with urea N applications at this rate. Since each 100 kg of soybean seed contains 6.5 kg of N, an application of 20 kg N supplies the N for only about 300 kg of beans (approximately 5 bushels per acre).

Garcia[21] emphasized that in his studies application rates of more than 20 kg of N per hectare at any one spraying resulted in serious leaf burn and should be avoided. Similar as the amount of N increased from 80 to 160 kg/ha. The application of 80 kg N per hectare required four sprayings with individual sprayings spaced at weekly intervals throughout the seed-filling period. The application of 80 kg N per hectare with P, K, and S resulted in a yield increase of 1040 kg/ha. In other experiments, applications of 120 kg N per hectare resulted in no yield increase.

D. Effect of Biuret and Other Urea Decomposition Products

Several substances may form in urea during storage due to the decomposition of urea.[21] Some of these decomposition products are potentially toxic to plants. Biuret has been shown to be an intermediate product in the formation of cyanate[22] and cyanamide[23] in urea; the formation of each of these is influenced by temperature and pH. Of these products, biuret has been studied most extensively and has been shown to be toxic to plants such as wheat[24] and rice.[23]

In a study of Meschi,[24] the effects of different concentrations of biuret, carbamate, cyanate, and cyanamide in NPKS solutions, with regent-grade urea as the source of N, sprayed on Amsoy and Hawkeye soybeans during the seed-filling period were compared. Before addition of urea to the nutrient solution, the urea solution was passed through Amberlite® MB-1, a mixed bed exchange resin, to remove any small amounts of cyanate present. Each of the urea decomposition products were included in the spray solutions as 0, 0.5, 1, 2, 4, and 8% of the total N. Beginning at growth stage R5, the soybean plants were sprayed four times at weekly intervals with an NPKS solution. The spray solutions contained N + P + K + S in a ratio of 10:1:3:0.5 and were sprayed on the plants in the evening to supply 20 kg N per hectare per spraying. Although leaf burn resulted if 2% or more of the N applied was any one of the above decomposition products, seed yields were reduced only where cyanate or cyanamide constituted 4% or more of the N applied. Similarly, Poole et al.[25] supplied an NPKS solution containing up to 4% of the N as biuret on soybeans at one or three times during the seed-filling period and did not reduce yields.

E. Effect of Chelating Agents

Although the monovalent cations such as K are readily retranslocated in plants, some nutrient elements, especially Ca but also Mn, Mg, and Fe, are not effectively retranslocated from the leaves and other plant parts to the developing seeds.[26] The concentrations of these elements (especially Ca as shown in Table 1) in the seeds are low and may limit seed development.[27]

Chelation of these di- and trivalent ions with organic acids that are phloem mobile in plants could potentially improve the translocation of these nutrient elements to the developing seeds.[28] Organic acids such as citric, malic, and malonic acids are naturally occurring chelates in plants. Johnson[29] reported that Ca chelates most strongly with citric acid, less strongly with malic acid, and only slightly with lactic or gluconic acid. Yazdi-Samadi et al.[30] reported the presence of malic and citric acids at low concentrations in the developing seeds of two soybean cultivars.

El-Hout[31] sprayed NPKS solutions containing lactic, citric, malic, or gluconic acid on Amsoy, Hark, and Harcor soybeans. The NPKS spray was applied four times during the seed-filling period for a total application of 80 + 8 + 24 + 4 kg/ha of N + P + K + S. The amounts of the acids applied per hectare were calculated to be equivalent to the amounts required to chelate 4 kg of Ca. Solutions were prepared so a spraying of 750 ℓ/ha applied the desired amounts of nutrients and chelating agents per spraying. Sprayings were made in the evening after 7:30 p.m. The NPKS foliar spraying resulted in a seed yield increase of 520 kg/ha over a check yield of 4370 kg/ha for the Amsoy variety, but no yield increase for the other varieties. None of the added organic acids had any effect on the yields of any cultivar and had no effect (except a decrease from 0.21 to 0.19% Ca in Amsoy) in the Ca concentration of the soybean used.

F. Effect of Solution pH

Barel and Black[32] applied orthopolyphosphate to soybean leaves at pH 2, 4, 5.5, 7, 8.5, and 10. More leaf burn was observed at pH 2 than at any higher pH.

Smith[12] applied NPKS solutions on soybeans at pH 2.7, 5.3, and 7.2 in 1977 and at pH 3.7 and 5.4 in 1978. Bean yields were not influenced significantly by the foliar applications in either year. Applications of solutions of pH 2.7 and 3.7 resulted in more leaf burn than resulted from solutions of higher pH.

Since ammonia can be lost from alkaline solutions when urea decomposes to release ammonia and since more leaf burn results from applications of more acidic solutions, the optimum pH range for foliar fertilizer solutions appears to be between pH 5 and 7.

G. Effect of Foliar NPKS on Nutrient Depletion and Photosynthetic Rate in Leaves

In research conducted in Iowa by Sesay and Shibles[11] and Smith,[12] foliar applications of NPKS solutions were applied in four sprayings during the seed-filling period to supply 80 + 8 + 24 + 4 kg/ha of N + P + K + S as described by Garcia and Hanway.[6] Bean yields were not increased by foliar fertilization in either study. The foliar fertilizer applications had no effect on the N content of the leaves but resulted in higher P and K contents of the leaves throughout the seed-filling period. The photosynthetic rates, soluble protein contents, and chlorophyll contents of the leaves declined markedly during the seed-filling period and were not affected by the foliar fertilizer applications.[11]

Studies in Florida by Boote et al.[10] produced similar results. The P and K concentrations in the leaves were maintained at higher levels by foliar fertilization, but the N concentrations and gross photosynthesis in the leaves were not influenced. Foliar fertilization did not significantly influence the seed-filling rate, number of seeds, or yield.

V. *Zea mays* (CORN)

The yield response of eight corn hybrids to NPKS foliar fertilization during the grain-filling period was studied by Kargbo.[33] Foliar applications of a fertilizer solution were made four times at 8-day intervals, beginning at 3 weeks after 75% silking. The fertilizer solution applied supplied a total of 60 + 12 + 6 kg/ha of N + P + K + S and was prepared using urea, potassium polyphosphate, and ammonium sulfate with 0.5% Tween® 80 (a surfactant). Spraying was done between 7 and 10 a.m. at a rate of 500 ℓ/ha. Of the eight hybrids tested, only one, a two-eared hybrid, showed a significant increase in grain yield — from 75.1 to 91.6 quintals/ha. Seed size was not affected by the foliar fertilization. In an adjoining experiment with one of the hybrids that showed no yield increase from the foliar fertilization there were indications that foliar applications later in the grain-filling period were more effective than applications earlier in the period.

Similar foliar applications were made by Snyder[34] and by Harder et al.[35] Snyder showed a small increase in grain yield from foliar fertilization when applications began 2 weeks after silking but no yield increase when applications began 4 weeks after silking. Harder et al. reported no significant effect of foliar fertilization on yield in 1 year when the applications began 4 weeks after silking and a significant reduction in yield another year when applications began 2 weeks after silking. Foliar fertilization resulted in significant increases in percentages of N and P and in the grain at harvest.

VI. SMALL GRAINS (OATS, RICE, WHEAT, AND BARLEY)

Oats *(Avena sativa)* — A field study of foliar NPKS fertilization of nine varieties of oats during the grain-filling period was conducted by Subah.[36] Applications consisted of four sprayings at 4, 11, 18, and 25 days after heading[37] or two sprayings at 4 and 11, 11 and

18, or 18 and 25 days after heading. Each spray application of 508 ℓ/ha supplied 22 + 4.5 + 3 + 2.4 kg of N + P + K + S per hectare. The fertilizer solution was prepared using urea, ammonium polyphosphate, potassium polyphosphate, and ammonium sulfate. Sprayings were made early in the morning or after sunset. Grain yields of the different varieties without foliar fertilization varied from 3850 to 4870 and averaged 4380 kg/ha. Yield increases of the different varieties resulting from four sprayings varied from 320 to 980 and averaged 730 kg/ha. Average grain yield increases resulting from two sprayings were not significantly different among different times of application and varied among varieties from 120 to 680 and averaged 350 kg/ha. Foliar fertilization resulted in increased numbers of harvested seeds but had no effect on seed size. The average % N in the harvested grain was increased by 0.34 to 0.42 over the average of 2.53 in the grain from unfertilized plots.

Rice (*Oryza sativa* L.) — Foliar fertilization of rice with NPKS solutions were studied for 3 years in Mississippi by Thom et al.[38] Fertilizer solutions were prepared using urea, potassium polyphosphate, and potassium sulfate. Yield increases of brown (unmilled) rice resulted in 1976 and 1978 but not in 1977. Rough rice yields of 5860 kg/ha in 1976 and 7086 kg/ha in 1978 were obtained without foliar fertilization. A foliar application of N + P + K + S at the rate of 54 + 12 + 22 + 2 at the midboot stage increased yields by 787 kg/ha in 1976 and at the rate of 54 + 8 + 15 + 2 at the panicle initiation and midboot stages increased yields by 1362 kg/ha in 1978. These foliar fertilizer applications increased % N in the brown rice 1.23 to 1.42 in 1976 and from 1.40 to 1.62 in 1978. Seed size was reduced from 22.7 to 18.2 g/1000 kernels by the foliar application in 1978.

Wheat (*Triticum aestivum* L.) — Foliar applications of solutions containing N, P, K, and S in the ratio of 20:2:6:1 applied in six sprayings supplying 30 kg N per hectare increased wheat yields in Australia in 2 years.[39]

VII. GENERAL DISCUSSION

The research that has been conducted indicates that foliar fertilization with several of the major nutrients during the seed-filling period offers a unique and real potential for increasing yields of the grain crops. But, it has not been consistently effective and there are problems to be solved. (I am reminded of the research on fertilizer use of 30 to 40 years ago when the first field experiments were being conducted which led to the development of the soil fertilization practices in common use today.)

The research that has been conducted on foliar fertilization provides leads concerning the forms and amounts of the different nutrients that can and should be applied and the times and methods of application. However, the amount of research that has been conducted on most of the grain crops is extremely limited, and much more research on soybeans will have to be conducted before foliar fertilization during seed-filling is likely to become a commonly used practice. Although there have been many failures, the successful studies indicate that there may be a real future for foliar fertilization. Studies such as those on the use of urease inhibitors,[40] synthetic cytokinins to delay leaf sensescence[41,42] and different surfactants may provide information that will lead to common, effective use of foliar fertilization for grain crops.

REFERENCES

1. **Boynton, D.,** Nutrition by foliar application, *Annu. Rev. Plant Physiol.,* 5, 31, 1954.
2. **Wittwer, S. H. and Bukovac, M. J.,** The uptake of nutrients through leaf surfaces, *Handb. Planzenernährung and düngung band,* 1, 235, 1969.

3. **Lawn, K. J. and Brun, W. A.**, Symbiotic nitrogen fixation in soybeans. I. Effect of photosynthetic source-sink manipulations, *Corp. Sci.*, 14, 11, 1974.

4. **Thibodeau, P. S. and Jaworski, E. G.**, Patterns of nitrogen utilization in the soybean, *Planta*, 127, 133, 1975.

5. **Hanway, J. J.**, Foliar fertilization of soybeans during seed-filling, in *World Soybean Research Conference II: Proceedings*, Corbin, F. T., Ed., Westview Press, Boulder, Colo., 1980, 409.

6. **Garcia, L. R. and Hanway, J. J.**, Foliar fertilization of soybeans during the seed-filling period, *Agron. J.*, 68, 653, 1976.

7. **Fehr, W. R. and Caviness, C. E.**, Stages of Soybean Development, Special Rep. 80, Iowa State University Coop. Ext. Serv. 1980.

8. **Vasilas, B. O., Legg, J. O., and Wolf, D. C.**, Foliar fertilization of soybeans: absorption and translocation of ^{15}N-labeled urea, Agron. J., 72, 271, 1980.

9. **Parker, M. B. and Boswell, F. C.**, Foliar injury, nutrient intake, and yield of soybeans as influenced by foliar fertilization, *Agron. J.*, 72, 110, 1980.

10. **Boote, K. J., Gallaher, R. N., Robertson, W. K., Hinson, K., and Hammond, L. C.**, Effect of foliar fertilization on photosynthesis, leaf nutrition and yield of soybeans, *Argon. J.*, 70, 787, 1978.

11. **Sesay, A. and Shibles, R.**, Mineral depletion and leaf senescence in soya bean as influenced by foliar nutrient application during seed filling, *Ann. Bot.*, 45, 70, 1980.

12. **Smith, T. L.**, The Effects of Solution pH and Nitrogen Rate on the Yield of Foliar Fertilized Soybeans, Ph.D. thesis, Iowa State University Library, Ames, 1980.

13. **El-Hout, N. M.**, The Effects of Chelating Agents in Foliar Fertilization Application on the Yield and Nutrient Content of Soybean Plants, M.S. thesis, Iowa State University, Library, Ames, 1982.

14. **Barel, D. and Black, C. A.**, Foliar application of P. I. Screening of various inorganic and organic P compounds, *Agron. J.*, 71, 15, 1979.

15. **Barel, D. and Black, C. A.**, Foliar application of P. II. Yield responses of corn and soybeans sprayed with various condensed phosphates and P-N compounds in greenhouse and field experiments, *Agron. J.*, 71, 21, 1979.

16. **Wittwer, S. H.**, Foliar absorption of plant nutrients, *Adv. Frontiers Plant Sci.*, 8, 161, 1964.

17. **Koonts, H. and Biddulph, O.**, Factors affecting absorption and translocation of foliar applied phosphorus, *Plant Physiol.*, 32, 463, 1957.

18. **Teubner, F. G., Wittwer, S. H., Long, W. G., and Tukey, H. B.**, Some factors affecting absorption and transport of foliar applied nutrients, *Mich. Agric. Exp. Stn. Q. Bull.*, 39, 398, 1957.

19. **Roesler, K. R.**, Effects of Rates and Timing of Foliar Fertilization on Yield Components and Nutrient Content of Soybeans, M.S. thesis, Iowa State University Library, Ames, 1982.

20. **Garcia, L. R.**, Foliar Fertilization of Soybeans During the Seed-Filling Period, Ph.D. thesis, Iowa State University Library, Ames, 1976.

21. **Welles, H. L., Giaquinto, A. R., and Lindstrom, R. E.**, Degradation of urea in concentrated aqueous solution, *J. Pharm. Sci.*, 60(8), 1212, 1971.

22. **May, D. R.**, Cyanamides, in *Kirk-Othmer Encyclopedia of Chemical Technology*, Vol. 7, 3rd ed., Grayson, M., Ed., John Wiley & Sons, New York, 1979, 291.

23. **Bhargara, B. S., Ghosh, A. B., and De, R.**, Effect of biuret concentration in foliar sprayed urea on yield and composition of rice, *Indian J. Agron.*, 29(1), 11, 1975.

24. **Meschi, M. A.**, The Toxicity of Urea Decomposition Products in Foliar Fertilizer Applied to Soybeans, M.S. thesis, Iowa State University Library, Ames, 1981.

25. **Poole, W. D., Randall, G. W., and Ham, G. E.**, Foliar fertilization of soybeans. II. Effect of biuret and application time of day, *Agron. J.*, 75, 201, 1983.

26. **Hanger, B. C.**, The movement of calcium in plants, *Comm. Soil Sci. Plant Anal.*, 10(162), 171, 1979.

27. **Hocking, P. J. and Pate, J. S.**, Mobilization of minerals to developing seeds of legume, *Ann. Bot.*, 41, 1259, 1977.

28. **Stewart, I.**, Chelation in the absorption and translocation of mineral elements, *Annu. Rev. of Plant Physiol.*, 14, 295, 1962.

29. **Johnson, H. W.**, Chelation between calcium and organic anions, *N.Z. J. Sci. Technol.*, 37B, 552, 1956.

30. **Yazdi-Samadt, B., Rinne, R. W., and Seif, R. D.**, Components of developing soybean seeds: oil, protein, sugars, starch, organic acids, and amino acids, *Agron. J.*, 69, 481, 1977.

31. **El-Hout, N. M.**, The Effects of Chelating Agents in Foliar Fertilizer Application on the Yield and Nutrient Content of Soybean Plants, M.S. thesis, Iowa State University Library, Ames, 1982.

32. **Barel, D. and Black, C. A.**, Effect of neutralization and addition of urea, sucrose, and various glycols on phosphorus absorption and leaf damage from foliar-applied phosphorus, *Plant Soil*, 52, 515, 1979.

33. **Kargbo, C. S.**, Foliar Fertilization of Corn During the Grain-Filling Period, Ph.D. thesis, Iowa State University Libray, Ames, 1978.

34. **Snyder, R. L.**, Yield and Yield Components of Corn and Soybeans as Influenced by Late-Season Foliar Fertilization and Soil-Moisture Stress, M.S. thesis, Iowa State University Library, Ames, 1978.

35. **Harder, H. J., Carlson, R. E., and Shaw, R. H.,** Corn grain yield and nutrient response to foliar fertilizer applied during grain fill, *Agron. J.,* 74, 106, 1982.
36. **Subah, J. Qwelibo,** Foliar Fertilization of Oats *(Avena sativa)* During the Grain-Filling Period, M.S. thesis, Iowa State University Library, Ames, 1978.
37. **Smith, D.,** Yield and chemical composition of oats for forage with advance in maturity, *Agron. J.,* 52, 637, 1960.
38. **Thom, W. O., Miller, T. C., and Bowman, D. H.,** Foliar fertilization of rice after midseason, *Agron. J.,* 73, 411, 1981.
39. **Ozanne, P. G. and Petch, A.,** The application of nutrients by foliar sprays to increase seed yields, Plant Nutrition 1978, Proc. 8th Int. Colloqium in Plant Analysis and Fertilizer Problems, Auchland, 1978.
40. **Mulvaney, R. L. and Bremner, J. M.,** Control of urea transformations in soils, in *Soil Biochemistry* Paul, E. A. and Ladd, J. N., Eds., Marcel Dekker, New York, 1981, chap. 4.
41. **Neumann, P. M. and Giskin, M.,** Late season for foliar fertilization of beans with NPKS: effects of cytokinins, calcium and spray frequency, *Comm. Soil Sci. Plant Anal.,* 10(3), 579, 1979.
42. **Noodén, L. D.,** Regulation of senescence, in *World Soybean Research Conference II; Proceedings,* Corbin, F. T., Ed., Westview Press, Boulder, Colo., 1980.

Chapter 7

REGULATING FLOWER PRODUCTION, SENESCENCE, SEED YIELD, AND OIL CONTENT OF CROP PLANTS BY GROWTH REGULATOR TREATMENTS*,**

C. Dean Dybing and Charles Lay

TABLE OF CONTENTS

*   Cooperative investigations of the U.S. Department of Agriculture, Agricultural Research Service, and the South Dakota Agricultural Experiment Station, Brookings, SD 47005. Journal Series No. 2000.
** This paper reports results of research only. Mention of a pesticide in this paper does not constitute recommendation by the USDA, nor does it imply registration under FIFRA. Mention of a trademark, proprietary product, or vendor does not constitute a guarantee or warranty of the product by the USDA and does not imply its approval to the exclusion of other products or vendors that may also be suitable.

I. ABBREVIATIONS

3-CPA — 2-(3-Chlorophenoxy)propionamide

GA_3 — Gibberellic acid

BA — 6-(Benzylamino)purine

Promalin® — BA + gibberellins A_4 and A_7 (Abbott Laboratories)

Accel® — N-(phenylmethyl)-9-(tetrahydro-2H-pyran-2-yl)-9H-purin-6-Amine (Shell Chemical Co.)

TIBA — 2,3,5-Triiodobenzoic acid

Ethephon — (2-Chloroethyl)phosphonic acid (Union Carbide Corp.)

Kinetin — 6-(Furfurylamino)purine

PP528 — Ethyl 5-(4-chlorophenyl)-2H-tetrazole-2-acetate (ICI United States, Inc.)

CCC — (2-Chloroethyl)trimethylammonium chloride

CF — Methyl 2-chloro-9-hydroxyfluorene-9-carboxylate; chlorflurenol (EM Industries, Inc.)

DCF — Methyl 2,7-dichloro-9-hydroxyfluorene-9-carboxylate; dichlorflurenol (EM Industries, Inc.)

nBF — n-Butyl 9-hydroxyfluorene-9-carboxylate; n-butyl fluorenol (EM Industries, Inc.)

Naptalam — N-1-naphthylphthalamic acid

Tween® 80 — Oxysorbic (20 POE) polyoxyethylene sorbitan monooleate (ICI United States, Inc.)

Mancozeb — Coordination product of zinc ion and manganeous ethylenebis(dithiocarbmate) (Rohm & Haas Co.)

II. INTRODUCTION

For a number of years, chemicals have been applied to plants to produce various effects. Well-known examples include fungicides to control diseases, herbicides to kill unwanted plants, and various chemicals to control tissue color, size, sugar content, sprouting, readiness to harvest, etc. Chemicals that are used to alter such physiological or morphological characteristics are called plant growth regulators (PGRs). They include hormones (growth-regulating compounds produced naturally by the plant, excluding inorganic and organic nutrients) and synthetic chemicals. The latter have been discovered by synthesizing chemicals with structures similar to the natural hormones and measuring the biological activity of by-products from the chemical and pharmaceutical industries. Both the synthetic and the natural types of PGR are effective at very low concentrations, and they are usually applied as foliar sprays. This combination of low rate and simple spray application makes PGR treatment a very economical way to modify plants. Since the 1940s, the number of chemicals used as PGRs, the kinds of growth response brought about by their application, and practical applications have all increased dramatically.

In spite of these advances, there has been only limited success in one important area. This is the long-sought goal of increasing seed yield of field crops by chemical treatment. Such a discovery, if it were to be achieved for a number of the major food crop plants, would have important advantages for present-day farming practice. Some of these include: (1) greater flexibility in management decisions, since chemical sprays could be applied whenever favorable weather coincided with proper crop growth stage; (2) breaking of relationships between characteristics that cannot now be dissociated by genetic methods, such as the association between high percent flower abortion and production of large numbers of flowers in crops like soybeans, or the inverse relationship between protein content and yield in wheat; (3) taking greater advantage of the enormous capacity of industry to discover new chemicals and applying that capacity to the solution of yield problems in crops where progress by breeding for some reason has been slowed; and (4) identifying new ways of improving plant morphology and physiology that will increase yield so that these improvements can be built into the next generations of crops through conventional genetic procedures.

Since it is still uncertain whether PGRs can be used to increase seed yield for crops like the cereals and oilseeds, research on the topic is actively continuing around the world. This paper describes a series of experiments based on the idea that the goal of increasing yield through chemical treatment is too broad and needs to be better focused to improve chances of success. That narrowing of scope was to be accomplished by breaking yield into its component parts and concentrating on finding chemicals that alter the components one at a time. Components of seed yield are

1. Number of fruits per unit of land area
2. Number of seeds per fruit
3. Seed size

The first two appear to be determined during flowering and the third during the period of seed growth. The initial goal of our research was to increase the first component, number of fruits per unit area, by using PGRs to increase number of flowers per area.

Flax (*Linum usitatissimum* L.) was chosen for the first studies because flower production appears closely related to seed yield.[1,2] Flax grown in the north central U.S. normally begins flowering late in June from a late-April seeding. The single flowering period lasts about 15 days but can be longer if conditions are favorable. When flowering stops, the plants normally senesce. However, many varieties can flower in cycles, i.e., produce flowers in two or more separate periods, in other environments.[1-4] Studies in the Imperial Valley of California have shown that yield of 3800 kg/ha or more can be obtained due in part to such cyclic flowering patterns.[2] Maximum commercial yields in north central states are about half those observed in California.[1] Physiological processes that cause flowering to occur in cycles and practical methods for altering individual flowering and rest periods are both unknown.[3]

After several years of tests on flax, the scope of our growth-regulator studies was expanded to include other goals and other crops. Added goals were (1) delay of senescence to prolong photosynthetic activity beyond the usual time of leaf death and (2) improvement of quality characteristics such as oil content. Crops tested in addition to flax were soybeans (*Glycine max* (L.) Merrill), sunflower (*Helianthus annuus* L.), wheat (*Triticum aestivum* (L.) Thell), and oats (*Avena sativa* L.). Only regulators showing some promise on flax were tested on the other crops.

III. GROWTH-REGUALTOR STUDIES

A. Growth Chamber and Laboratory Studies of Flax Flowering

For studies of flax flowering, we classified buds according to location and activity (Table 1).[5] Cotyledonary buds produced branches during vegetative growth when soil moisture levels were adequate. Upper mainstem buds produced panicle branches during reproductive growth. Other mainstem buds were usually inactive unless the plant was lodged. Panicle buds consisted of terminal buds on the branches plus buds in the axils of bracts.

Study of bud groups during flowering showed that all had been initiated to flower, but many failed to produce blossoms.[1,3,5] Thus, terminal buds were active throughout all blossoming cycles, but axillary buds were mostly inactive unless stimulated to shoot and flower production by continuous excision of all flowers prior to fruit growth. Removal of all fruits at the end of the first flowering period stimulated new flowering activity at terminal buds but not at axillary or mainstem buds. All buds, panicle and mainstem alike, readily produced flowers when grown in vitro if hormones were omitted from the culture media.[3] Taken together, these findings were thought to mean that chemical inhibition occurs in mainstem

Table 1

SHOOT GROWTH AND FLOWER PRODUCTION ACTIVITIES FROM FLAX BUDS AT DIFFERENT LOCATIONS ON THE PLANT

| | Activity during flowering | | Activity in response to various treatments[a] | | | | | | |
| | | | Level of Activity[b] | | | Foliar applications | | | |
Bud type	First period	Later periods	High moisture	Flower removal	In vitro culture	Auxin	Gibberellin	Cytokinin	Antiauxin
Main stem									
Cotyledonary node	0	0	+++	+	—[c]	0	0	0	+++
Middle nodes	0	0	0	0	+++	0	+	0	+++
Upper nodes	+	+++	++	++	+++	0	++	0	+++
Panicle									
Axillary	+	+++	++	++	+++	+	0	++	+++
Terminal	+++	+++	+++	++	+++	0	0	0	0

[a] Treatments: High moisture = irrigation or extensive inter-plant spacing; flower removal = continuous excision of new flowers; in vitro culture = agar-based organ culture of single node with leaf and bud; foliar applications = IAA, GA₃, kinetin, or 2,3,5-triiodobenzoic acid antiauxin at rates of 250 to 500 ppm, applied at end of first flowering period. Variety = CI 1303.

[b] 0 = none; + = slow shoot growth; + + = moderate shoot growth and flower production; + + + = extensive shoot growth and flower production.

[c] Not tested.

buds, in many auxiliary panicle buds during each flowering period, and in terminal branch buds in rest periods between periods of flowering.

Assuming a chemical inhibition of the various kinds of buds, we thought that treatment with the proper chemicals might counteract the inhibition, stimulate growth from the buds, permit additional flower and fruit development, and thereby increase seed yield. Therefore, trials were conducted where foliar sprays were applied to intact plants in the growth chamber. Plant growth regulators were indeed seen to influence the activities of the various types of bud (Table 1).[5] Auxins, especially the phenoxy auxins, tended to inhibit shoot and flower development at all buds. Gibberellin inhibited panicle buds but promoted shoot growth at stem buds. Kinetin was generally inhibitory at high concentration but stimulated growth at axillary panicle buds at low concentration. The antiauxin TIBA stimulated shoot and flower development from all except the terminal buds.

B. Field Growth-Regulator Trials with Flax

Since growth chamber and laboratory tests showed that some growth regulators could release inhibited buds and produce additional flowers, field testing of regulators was started.[6] Little published information on effects of growth regulators on flax was available, so the first tests were designed to screen numerous compounds in each of several regulator types. Field testing was initiated in 1973 and completed in 1983. In all, 57 chemicals were evaluated. Data will be reported here only for representatives of principal regulator classes and, except where stated otherwise, for the flax variety 'Linott'.

Methods and experimental designs used in this work have been described in detail elsewhere.[7-10] All tests were conducted at Brookings, S.D. The plots generally were 4.3 to 6.9 m^2 in area and had four rows with 0.36 m row spacing. Seeding was in late April to early May; other procedures were typical of commercial or field plot practice. Regulators were applied with a four-nozzle knapsack sprayer operating at 1.1 kg/cm^2 pressure and applying 280 to 350 ℓ/ha. Tween® 80 was included in the solutions as surfactant at a concentration of 0.075% (w/v). Applications to flax were made in vegetative, bud, full bloom, and postbloom growth stages. Individual regulators were applied only once to any plot in an experiment, and some regulators were tested only a few times.

In the first 3 years of the program, most of the growth effects that were observed either were unrelated to seed yield or even tended to decrease it.[6,11] Plant height was increased by GA_3 applied in bud stage and reduced by retardants like CCC applied in vegetative stage; neither effect significantly increased yield. Branching at the cotyledonary node was stimulated by ethephon, TIBA, or CF aplied in the vegetative stage (Table 2), but associated effects like fascination of upper leaves and epinasty of the stems interfered with panicle development. Flowers, fruits, and seeds were deformed by several chemicals when applications were made in bud or bloom stages.

On the other hand, one effect that appeared more promising was senescence delay caused by chemicals applied "postbloom", i.e., at the end of the first flowering period. In tests through 1975, both GA_3 and the cytokinin Accel® delayed plant death without detrimental side effects when applied postbloom.[11] This effect was evidenced as temporary delay in loss of canopy chlorophyll. It appeared to be a natural outcome of bud release, since the growth resulting from stimulated bud activity included shoots with new leaves, bracts, and sepals. Although the delay in plant death observed up to 1976 was not long enough for extensive new flower production, more pronounced delay appeared likely if chemicals with greater activity could be found. Antiauxins and morphactins appeared especially promising in this regard.[11] Delay of senescence was therefore emphasized as a program goal beginning in 1976, with senescence defined as canopy yellowing and estimated visually on a scale of 1 to 10, or by measurement of extractable chlorophyll. Leaf area and renewed flowering were also recorded. Measurement of leaf protein, metabolic activity, or other characteristics that

Table 2

GROWTH REGULATOR EFFECTS ON FIELD-GROWN LINOTT FLAX

Representative			Growth regulator effect[a]							
				Senescence delay		Seed yield		Seed oil %		
Regulator class	Chemical	Rate (g/ha)	Bud release	Slight	Pronounced	Decreased	Increased	Decreased	Increased	
Auxin	3-CPA	200	4	4	3	2,3,4	—	4	—	
Gibberellin	GA₃	300	—	3	—	2,3	—	1,4	—	
Cytokinin	Accel®	200	4	—	4	2	—	—	4	
Ethylene stimulator	Ethephon	500	1,4	4	3	1,2,3,4	—	1,3,4	—	
Retardant[b]	CCC	4000	—	1,2,4	—	—	—	—	—	
Antiauxin	TIBA	100	1,2,3,4	3	3	1	3	1	4	
Morphactin	CF	25	1,2,3,4	—	1,2,3,4	1,2,3	4	1	4	

[a] 1 = vegetative stage; 2 = bud stage; 3 = bloom stage; 4 = postbloom; — means effect not observed.

[b] Retardants applied in vegetative stage only.

Table 3
GROWTH REGULATOR EFFECTS ON CHLOROPHYLL RETENTION AND NEW SHOOT PRODUCTION BY LINOTT FLAX (1976 EXPERIMENT)

Treatments					New shoots[b] (No./ 0.03m^2)	
Application date	Regulator	Rate (g/ha)	Visual rating[a]	% Green stems[b]	Panicle	Main stem
Vegetative	3-CPA	200	1.6	—	—	—
	GA$_3$	200	3.3	—	—	—
	Ethephon	900	1.6	6	1	1
	TIBA	100	2.7	5	2	0
	PP528	600	10.0**	28*	9	0
	CF	25	10.0**	4	0	1
Bloom	3-CPA	200	6.6**	6	1	2
	GA$_3$	200	3.3	—	—	—
	Ethephon	900	9.4**	12	2	2
	TIBA	100	3.8*	24	4	20**
	PP528	600	7.8**	41**	14**	18**
	CF	25	6.6**	79**	30**	15**
Postbloom	3-CPA	200	1.0	9	2	5
	GA$_3$	200	1.6	—	—	—
	Ethephon	900	2.1	24	8	12**
	TIBA	100	2.7	10	2	7
	PP528	600	2.1	17	5	13**
	CF	25	7.2**	53**	19**	23**
—	Check	—	1.0	10	4	2

Note: *, **Significantly different from the check at 0.05 and 0.01 levels of probability, respectively.

[a] 1 to 10 scale; 1 = 1 to 10% and 10 = 91 to 100% apparent canopy chlorophyll retention, 27 days after full bloom.
[b] 55 days after full bloom.

would have characterized senescence more directly than canopy yellowing[12] was not attempted because of the large number of plots involved in experiments evaluating numerous chemicals at several rates and dates of application.

C. Field Studies of Senescence Delay and Seed Yield of Flax

From 1976 on, senescence delay was observed in every year of testing with flax (Table 3).[7,9] 11 regulators tested 2 or more years gave significant delay in at least 1 year. CF, DCF, PP528, and TIBA were consistently effective. *n*-Butyl flurenol was sometimes effective, but the response was less pronounced and less consistent than that observed with the other flurenol compounds. Ethephon and 3-CPA gave senescence delay in 2 of 3 years when applied in bloom and postbloom stages. The cytokinins, Accel® and Promalin®, gave significant response in one trial each, but BA applied alone was ineffective. Other chemicals effective in one trial each were GA$_3$ and naptalam.

Growth stage at the time of regulator application had a marked influence on expression of senescence (Table 3). The maximum delay of 56 days was observed in 1976 for PP528 applied in the bloom stage and chlorflurenol applied in the bud or bloom stages. For both treatments, the delay appeared due to abnormalities in growth and morphology, since they caused failure in fruit set as well as prolonged leaf retention. Postbloom applications, as observed in earlier studies, caused fewer growth abnormalities. The maximum delay in

Table 4
EFFECTS OF CHLORFLURENOL ON SEED YIELD AND SEED OIL CONTENT

Crop	Trial	Year	Treatment Date[a]	Rate (kg/ha)	Treated as % of check Seed yield	Oil content
Flax	1	1976	Postbloom	25	115*	112**
(Linott)	2	1977	Postbloom	25	94*	108**
	3	1977	Postbloom	25	82*	103**
	4	1978	Postbloom	25	87*	104**
	5	1978	Postbloom	25	96	104**
	6	1979	Postbloom	25	110*	104**
	7	1980	Postbloom	25	94	106**
	8	1983	Postbloom	25	103	106**
Soybeans	1	1978	R6	25	91**	108**
(Swift)	2	1979	R6	25	95	101**
	3	1980	R6	25	97	102**
	4	1983	R4	25	87	102
Wheat	1	1978	Postanthesis	500	127*	—
(Era)	2	1979	Postanthesis	500	113*	100
	1	1978	Postanthesis	1000	97	—
	2	1979	Postanthesis	1000	105	102*
Oats	1	1979	Postanthesis	500	99	102*
(Dal)	1	1979	Postanthesis	1000	97	104**
Sunflowers	1	1980	Postanthesis	50[b]	—	97
(Hybrid 894)						

Note: *, ** Treated is significantly different from the check at 0.05 and 0.01 levels of probability, respectively.

[a]   R6 = full seed growth stage for soybean; wheat and oats treated 2 to 3 days after anthesis; sunflowers treated 1 day after anthesis of innermost ring of flowers.
[b]   50 ppm applied to runoff.

harvest date from a postbloom treatment was 36 days for chlorflurenol in 1976. CF applied postbloom increased chlorophyll content of individual leaves and quantity of leaf tissue per plot.[7] New shoot growth was stimulated from both panicle and mainstem buds (Table 3).

Although the shoot growth stimulated by CF and other chemicals applied postbloom caused early resumption of flowering and production of additional fruits, there was no consistent relationship between this "senescence delay" and yield (Table 4). Regulators 3-CPA, Promalin®, ethephon, and PP528 retarded chlorophyll loss but also tended to reduce yield. TIBA and Accel® delayed yellowing without a significant effect on yield. CF consistently delayed senescence, but at the rate of 25 g/ha it gave inconsistent yield response (Table 4). DCF significantly increased yield in 1977.

To further evaluate these variations in response, flax yield was divided into its components within pre- and postspray periods in two field experiments.[8] CF applied postbloom in the field under rainfed conditions did not affect fruits formed from flowers that had reached anthesis before regulator treatment, but it did stimulate secondary fruiting and added an additional 34 bolls per 0.03 m^2 or 23% of boll production in the prespray flowering period. However, yield was not increased because the new bolls had only 1.7 seeds per fruit on the average and seeds were about half as large as those produced in the prespray period. Under irrigated conditions, postspray flowering was relatively high in both treated and untreated plots, and again seed yields were not significantly increased by the spray.

D. Field Studies with Soybeans, Sunflowers, Wheat, and Oats

Experiments with soybeans employed a Group O cultivar 'Swift'.[7,8] Chlorflurenol applied at R5 stage at rates of 25 or 50 g/ha visibly delayed yellowing of Swift both in 1978 and 1979. Applications at R6 was effective in 1978 but not 1979. Senescence delay was also observed in 1978 at rates of 12.5, 75, 100 g/ha. Growth abnormalities accompanied senescence delay at 100 g/ha.

As in the case of flax, senescence delay of soybean was not associated with increased seed yield. CF significantly decreased yield due to a reduction in seed size when applied at 25 and 50 g/ha at the R3 through R7 stages (Table 4). Treatment with nBF at R6 stage in 1979 increased yield 14% but had no effect on leaf yellowing.

For the cereals, CF was applied at boot, anthesis, and three postanthesis stages at rates ranging from 500 to 2000 g/ha on 'Era' wheat and 500 to 1000 g/ha on 'Dal' oats.[7,8] Other chemicals tested were DCF and nBF at 1000 g/ha, and BA at 500 g/ha. Reduction of both height and lodging were observed, but none of the treatments retarded senescence of either crop. Postanthesis CF treatments at the 500 g/ha rate increased seed yield of wheat in 1978 and 1979 (Table 4), and DCF increased yield of oats when applied in boot stage. CF applied in boot stage decreased oat seed size, and, consequently, decreased yield as well.

Sunflowers were evaluated in only 1 year (1980), and they were the only crop tested that did not respond to chlorflurenol.[10] The regulator was applied to runoff as a foliar spray at rates up to 50 ppm. No other chemicals were tested on sunflower. Leaf chlorophyll content, achene weight, and percent seed set in outer, middle, and inner thirds of individuals heads were not significantly affected by CF treatment.

E. Effects on Oil and Protein Content of Seeds

Effects of growth regulators on quality characteristics are also important because flax, soybeans, and sunflowers are grown, in part, for their oil. CF applied postbloom to flax consistently increased percent oil (Table 4) and decreased protein.[9,10] As oil percent was increased, oil yield (kilograms oil per hectare) tended to rise and protein yield tended to decline. Oil yield was highly correlated to seed yield (r = 0.99**). DCF and TIBA also increased oil content, but the other regulators tested either decreased oil or had no effect (Tables 2 and 5). Iodine value, a measure of drying quality of oil, was not significantly affected by regulator treatments.

In 3 of 4 years of testing on soybeans, CF significantly increased percent oil when applied at the rate of 25 g/ha in R6 growth stage (Table 4). As with flax, the increase in the oil content of soybean was accompanied by a decrease in the percent protein (r = − 0.87** in 1978). Significant effects on seed oil content were likewise detected for wheat and oats treated with chlorflurenol but not sunflowers. Extent of the change in oil content for all these crops was relatively small compared to the effect on flax. DCF and nBF were less effective than CF.

Since the two chemicals showing the greatest effect on seed oil and protein also exhibited a tendency to alter seed set and seed size if applied during flowering, it is possible that changes in chemical composition may be due to incomplete seed development. To test this possibility, relationships of seed oil and seed size were determined. Seed size in flax is expressed as weight per 1000 seeds. For trials involving morphactins, from 1975 to 1983, increase in oil percent was often accompanied by a decrease in seed size (Table 5). However, probability of an increase in oil content without a decrease in seed size was high if rate of application was low (i.e., 6 g/ha for CF and 50 g/ha or less for DCF). Delaying applications to later dates also decreased the likelihood of a reduction in seed size.

Table 5
RELATIONSHIP BETWEEN INCREASE IN OIL
CONTENT AND DECREASE IN SEED SIZE FOR
LINOTT FLAX TREATED POSTBLOOM WITH
CHLORFLURENOL

			Frequency (% of trials)	
Regulator	Rate (g/ha)	No. trials[a]	Oil content increased	Seed size decreased
Chlorflurenol	6	3	67	0
	13	20	80	90
	25	45	87	75
	50	24	88	79
	100	7	100	100
Dichlorflurenol	25	2	100	0
	50	24	46	8
	100	27	51	46
	200	16	81	62

[a]   Trial = all replications of a chemical treatment in 1 experiment.
Data from all 1975 to 1983 field tests for applications from 9 to
33 days after full bloom.

F. Analysis of Problems Experienced in these Growth Regulator Studies
1. Variability in Yield Response

Although morphological effects of growth-regulator treatment were observed in every trial except the sunflower study, effects on yield were highly variable. For flax, yield was significantly increased by postbloom application of chorflurenol in 2 years, significantly decreased in 2 years, and unaffected in 4 years (Table 4). Similar variation in yield effect was observed with other regulators on flax and most regulators tested on soybeans and the cereals. Moreover, some cultivars of flax were more responsive than others.[8] Such inconsistency in yield enhancement from growth-regulator treatments has been observed by other workers on other field crops.[13,14] The cause of this variability of results in our trials is unclear. Some possible factors include inadequate penetration, incorrect timing of treatments, preharvest fruit drop, low seed set in new fruits, and uncontrolled disease conditions.

A possible explanation for variable effectiveness that must always be considered is failure of the chemical to penetrate the leaves in sufficient quantity to promote a response. We at first believed that sunflowers failed to respond to CF in our 1980 field trial because of inadequate uptake of the chemical. However, recent growth-chamber studies indicate that sunflower may be less responsive than soybeans, probably for reasons other than lack of penetration.[15] As for the other crops, we always tried to maximize uptake by including an effective surfactant (Tween® 80) in all spray solutions and by making applications under the best environmental conditions possible. Since every experiment showed morphological and/or oil content responses for at least some of the chemicals, and since highly effective chemicals like CF never failed to cause some morphological response, we believe that variability in uptake is not a likely reason for the variability in yield response that we observed. This contention is further supported by the observation that temporary epinastic response to CF was always detected a few hours after application, even where rainfall that might have washed unabsorbed chemical from the leaves, had occurred after treatment.

Improper timing of the treatments with respect to physiological age is also a factor that might contribute to variability of response. In this work, timing of treatments appeared to be critical, with too early an application causing detrimental side effects that reduced yield, and too late a treatment failing to increase yield because flowering was stimulated at a time of high temperature and moisture shortage. Chlorflurenol delayed leaf yellowing and stimulated flowering in all flax trials. However, applications during vegetative, bud, and bloom stages induced fruiting abnormalities, and these appeared to be related to the delay in yellowing[16] and also to reductions in yield. Withholding regulator treatments until termination of the first flowering period ("postbloom" treatment) favored senescence delay without detrimental effect on fruiting. Yet, the treatments had to be applied early enough in the life cycle to stimulate activity of panicle buds before irreversible senescence had been triggered by hot, dry summer weather. Solution of this problem may require a "slow release" system that would permit early treatment (bloom stage or earlier) but delay activity until a time when bud release and senescence delay could occur without damage. Such a system will probably require modification of either the active ingredient or the formulation.

Another factor to be considered is differential ability of plants to retain fruits for long periods beyond the normal harvest data. Since a delay in loss of canopy chlorophyll and moisture content also means delay of harvest date, it is necessary that the plants not shed existing mature fruits during the period new fruits are developed. Not all flax varieties tested were able to retain fruits adequately, so premature fruit drop accounted for most of the varietal differences that we observed. However, of the varieties tested, Linott, the variety used for all of the data reported here, appeared to have maximum ability to retain fruits. Year-to-year variation in yield response of Linott therefore does not appear to have been caused by excessive preharvest fruit shed.

As to seed production, secondary fruiting stimulated by chlorflurenol was not very efficient. Additional fruits were produced in every trial with CF, but low seed set and small seed size in the new fruits, plus reduced seed size for fruits set prior to treatment, presumably prevented a yield increase in some trials. Both environmental and chemical factors may have contributed to this low fecundity. Fruits added to the plants after treatment may have experienced moisture stress and elevated air temperatures since they were developing late in the growing season. Such stresses clearly reduce seed set and seed size.[17,18] However, environment was not always a limiting factor, because yield was not increased by CF in the one irrigated trial that we conducted. The chemical factor would be the interference with normal seed and fruit development mentioned previously. We attempted to avoid this effect, even for fruits formed after treatment, by keeping application rates low and searching for less detrimental chemicals. Rates of CF greater than 6 g/ha showed great likelihood for reducing seed size (Table 5), but rates lower than 25 g/ha were decreased in ability to delay senescence. The morphactins DCF and nBF and the antiauxin TIBA gave less fruiting abnormality than CF, but they also gave less senescence delay. Antiauxins having lesser biological activity than TIBA (naptalam, 2,4,6-trichlorophenoxyacetic acid, and others) showed little promise for senescence delay and yield enhancement at any stage of plant growth.

Finally, interaction of disease and senescence must be considered in evaluating the yield effects. Fungicide (Mancozeb) and CF both delayed canopy yellowing for flax. Additive effects of the two treatments were observed on flax yield (Table 6).[8] Pasmo[19] (*Septoria linicola* [Speg.] Gar.) and other foliar diseases that increase on flax late in the growing season may have reduced the photosynthetic capabilities of new leaves on shoots produced as a result of CF stimulation and old leaves retained beyond their normal life span. Thus, some CF treatments probably failed to increase yield because, in the absence of an accompanying fungicide treatment, disease conditions increased rapidly to eliminate any beneficial effect that the regulator might have.

2. Practical Application of the Oil Content Effect
One growth-regulator effect that was observed consistently throughout all years of the

Table 6
**ADDITIVITY OF FUNGICIDE
AND CHLORFLURENOL
EFFECTS ON SEED YIELD**

Treatment[a]	Seed yield (treated as % of check)		
	Flax	Wheat	Oats
Mancozeb	113	112	114
CF	106	118	103
CF + Mancozeb	127	116	117
LSD0.05	12	12	10

[a]   1979 trials; 900 g/ha Mancozeb applied
weekly from first bloom of flax, boot stage
of wheat, and anthesis of oats; CF applied
postbloom at 25 g/ha for flax, and 12 days
postanthesis at 1000 g/ha for wheat and oats.

program was increased oil content of seeds from plants treated with morphactins, particularly CF. A number of uncertainties must be answered before commercial practice can be developed from this effect. Since treatments must be made late in the growing season to avoid side effects deleterious to yield, aerial application of the chemicals would be required to minimize plant damage from spray equipment. Benefit of the oil change to the crushing industry will depend on magnitude of the increases in oil percent and oil yield per acre. Value to the grower, on the other hand, will depend more on effects on seed yield, unless a premium price is paid for higher oil content. A slow-release formulation of the chemical might help to solve the problem of growth stage at application, but much work remains to be done before details of the economic questions are clarified.

3. Is the Morphactin Effect Really a Delay of "Senescence"?
Throughout this work, senescence was defined primarily as canopy yellowing. Of course, senescence involves processes over and above loss of chlorophyll.[12,20] The fact that plants treated with morphactins become greener has long been recognized, and the effect has been ascribed to tight packing of chloroplasts in expanding leaves and retarded chlorophyll breakdown in aging leaves.[21] Our own studies have shown, however, that a close association exists between morphactin effects on chlorophyll and promotion of growth in the affected leaf.[22] For intact soybean plants treated with CF, rapid changes in chlorophyll content occurred, accompanied by significant increases in fresh and dry weight (Table 7). Changes in leaf density resulted from extensive cell division and elongation in mesophyll layers plus renewed activity of vascular cambium cells. Such effects on CF on soybean appear quite similar to "regreening" effects observed in decapitated plants where senescing lower leaves tend to recover and begin synthesis of chlorophyll, protein, and RNA.[23]
Even if the morphactins directly reduce chlorophyll breakdown in senescing leaves, the effect appears to vary by crop species. Wheat and oats in the field were affected by CF both morphologically and in oil content of the seeds, but they were not delayed in leaf yellowing.[7] Leaf disks of dock (*Rumex obtusifolius* L.) and radish (*Raphanus sativus* L.) floating on chlorflurenol solutions in the dark under laboratory conditions were delayed in chlorophyll loss by CF, but disks of soybean, tobacco (*Nicotiana tabacum* L.), and barley (*Hordeum vulgare* L.) were not (Table 7).[22]
The effect of morphactins on oil content could also be interpreted as an effect on senes-

Table 7

CHLORFLURENOL EFFECTS ON INTACT PLANTS AND ON FLOATING LEAF DISKS

| | Intact soybean plants treated with CF (10^{-5} M) | | | | Floating leaf disks treated with CF (10^{-4} M) | |
| | Treatment period (days) | | Treated as % of check | | | Treated as % of check |
Tissue	Light	Dark	Chlorophyll	Weight[a]	Plant[b]	Chlorophyll
Cotyledon	8	0	192	542	Soybean	72
Unifoliate leaf	9	0	157	147	Dock	300
	5	4	432	—	Radish	222
Trifoliate leaf	9	0	148	128	Barley	108
	5	4	917	112	Tobacco	54

[a] Fresh weight of whole cotyledons or dry weight of 1.5 cm (diam) leaf disks.
[b] Leaf disks floating in the dark for 3 to 12 days on solution ± CF.

Table 8

INCORPORATION OF ACETATE-C^{14} INTO LIPIDS OF LINOTT FLAX EMBROYS

| | dpm/μg lipid | |
Treatment	Mean	s_x
Check	99.2 ± 29.9	
0.1 ppm chlorflurenol	125.0 ± 20.2	
0.5 ppm chlorflurenol	144.6 ± 24.9	
1.0 ppm chlorflurenol	207.9 ± 30.5	
5.0 ppm chlorflurenol	133.2 ± 39.9	
LSD0.10	77.6	

cence. That is, delay in leaf yellowing (plus addition of new leaves from stimulated shoot growth) could be assumed to increase photosynthetic activity and thus provide more energy to developing seeds for oil synthesis. We attempted to evaluate this possibility in preliminary studies of flaxseed lipid synthesis using 16 ± 2 day-old embryos incubated with acetate-C^{14} ± CF for 48 hr at room temperature (Table 8).[24] For each increase in CF concentration there was a corresponding increase in counts per microgram up to 1.0 ppm. The decrease in counts from 5.0 ppm was probably due to a toxic effect on the embryo from chlorflurenol. These data suggest a direct interaction of the morphactin with lipid metabolism in developing embryos in flax.

We conclude from these observations that effects of CF on crops like flax and soybean principally involves growth processes such as (1) addition of new leaves through bud release and shoot growth, (2) increased density through elongation and renewed cell division in mesophyll layers for leaves existing at time of spraying, and (3) direct effect on anabolic activities like lipid synthesis. Senescence delay occurs, in the sense that plant death and date of harvest are delayed. However, delay of chlorophyll degradation, as measured by leaf disks floating on solutions in the dark, would appear less important.

IV. FUTURE DEVELOPMENTS

These studies started with the idea that yield increases might be obtained through increased flower production resulting from growth regulator treatments. They continued with studies of senescence and lipid metabolism. The most promising outcome of the work was a consistently observed increase in seed-oil content. The original goal of increasing yield by increasing one of the components of yield remains valid but presently unattainable because of the variability in yield response. The multigenic regulation of seed yield, the compensation that causes one yield component to decline whenever another is increased, and the high likelihood of detrimental side effects from chemical treatments combine to make yield increase through growth-regulator treatment very difficult to obtain. Senescence under field conditions, particularly where disease is a factor, may be quite unlike the processes studied under growth chamber or Petri dish conditions. Physiological factors that regulate seed yield, whether nutritional, hormonal, or some other, are still only poorly defined and understood. However, growth regulators may yet contribute to improved levels of production of field crops, but future developments will depend heavily on improved knowledge and understanding of regulator molecules and the plant processes targeted for improvement by chemical treatment. Future research on growth regulators must include the continued search for more effective compounds and better ways of application. Research on plant processes should focus on basic mechanisms that regulate flowering, fruiting, seed set, and seed development.

ACKNOWLEDGMENT

We gratefully acknowledge the technical assistance of R. A. Carsrud and D. A. Roden, Agricultural Research Technicians, USDA-ARS. Gifts of plant growth regulators from companies mentioned under ''Materials and Methods'' are also acknowledged.

REFERENCES

1. **Dybing, C. D. and Carsrud, R. A.,** Flax: breaking the yield barrier, *S.D. Agric. Exp. Stn. Farm Home Res.*, 25(2), 23, 1974.
2. **Knowles, P. R., Isom, W. H., and Worker, G. F.,** Flax production in the Imperial Valley, *Calif. Agric. Exp. Stn. Circ.*, 480, 3, 1959.
3. **Hovland, A. S. and Dybing, C. D.,** Cyclic flowering patterns in flax as influenced by environment and plant growth regulators, *Crop Sci.*, 13, 380, 1973.
4. **Dybing, C. D. and Lay, C.,** Flax, in *CRC Handbook of Biosolar Resources*, Vol. II, McClure, T. A. and Lipinsky, E. S., Eds., CRC Press, Boca Raton, Fla., 1981, 71.
5. **Dybing, C. D.,** Progress report: regulation of flower bud development in flax, *Flax Inst. U.S. Proc.*, 42, 15, 1973.
6. **Dybing, C. D.,** Growth regulator trials with flax in 1973 and 1974, *Flax Inst. U.S. Proc.*, 44, 5, 1974.
7. **Dybing, C. D. and Lay, C.,** Field evaluations of morphactins and other growth regulators for senescence delay of flax, soybean, wheat and oats, *Crop. Sci.*, 21, 879, 1981.
8. **Dybing, C. D. and Lay, C.,** Yields and yield components of flax, soybean, wheat and oats treated with morphactins and other growth regulators for senescence delay, *Crop Sci.*, 21, 904, 1981.
9. **Dybing, C. D. and Lay, C.,** Oil and protein in field crops treated with morphactins and other growth regulators for senescence delay, *Crop. Sci.*, 22, 1054, 1982.
10. **Dybing, C. D. and Lay, C.,** Growth regulators that alter oil content and quality, *Plant Growth Regulator Bull.*, 10(3), 9, 1982.
11. **Dybing, C. D.,** Delayed senescence of flax treated with morphactins or anti-auxins, *Proc. Plant Growth Regulator Working Group*, 4, 207, 1977.
12. **Patterson, T. G., Moss, D. N., and Brun, W. A.,** Enzymatic changes during senescence of field-grown wheat, *Crop. Sci.*, 20, 15, 1980.

13. **Hardy, R. W. F.**, Chemical plant growth regulation in world agriculture, in *Plant Regulation and World Agriculture*, Scott, T. K., Ed., Plenum Press, N.Y., 1979, 165.
14. **Ries, S., Wert, V., and Biernbaum, J. A.**, Factors altering response of plants to triacontanol, *J. Am. Soc. Hortic. Sci.*, 108, 917, 1983.
15. **Dybing, C. D.**, unpublished data, 1984.
16. **Leopold, A. C., Niedergang-Kamien, E., and Janick, J.**, Experimental modification of plant senescence, *Plant Physiol.*, 34, 570, 1959.
17. **Dybing, C. D. and Zimmerman, D. C.**, Temperature effects on flax (*Linum usitatissimum* L.) growth, seed production, and oil quality in controlled environments, *Crop Sci.*, 5, 184, 1965.
18. **Dybing, C. D.**, Maturity and yield of seedflax in controlled environments: effects of root environment, *Crop Sci.*, 9, 572, 1969.
19. **Sackston, W. E.**, Effect of pasmo disease on seed yield and thousand kernel weight of flax, *Can. J. Res. Sect. C*, 28, 493, 1950.
20. **Lauriere, C.**, Enzymes and leaf senescence, *Physiol. Veg.*, 21, 1159, 1983.
21. **Schneider, G.**, Morphactins: physiology and performance, *Annu. Rev. Plant Physiol.*, 21, 499, 1980.
22. **Dybing, C. D. and Yarrow, G. L.**, Morphactin effects on soybean leaf anatomy and chlorophyll content, *J. Plant Growth Regulation*, 3, 9, 1984.
23. **Thimann, K. V.**, The senescence of leaves, in *Senescence in Plants*, Thimann, K. V., Ed., CRC Press, Boca Raton, Fla., 1980, 85.
24. **Dybing, C. D. and Lay, C.**, Effects of plant growth regulators which delay senescence on seed and oil yields in oilseeds, *Proc. Plant Growth Regulator Soc. Am.*, 10, 270, 1983.

Chapter 8

APPLICATION OF ETHYLENE-RELEASING COMPOUNDS IN AGRICULTURE

R. M. Beaudry and S. J. Kays

TABLE OF CONTENTS

I. INTRODUCTION

Ethylene, unlike other plant hormomes, is a gas at physiological temperatures and, as a consequence, has relatively unique requirements for agricultural use. It is known to promote, inhibit, or otherwise modulate a number of basic metabolic processes within plants. These, in turn, elicit a diverse array of physiological responses (Table 1), a number of which are of considerable agricultural interest (Table 2). Of particular interest has been the relationship between ethylene and abscission, fruit ripening, and senescence. Pre- and postharvest application of ethylene has allowed us to alter the normal timing of specific physiological events and/or their natural outcome to our advantage. Of the currently known hormones in plants, ethylene is the most widely and successfully used in agriculture. Its use represents a multimillion dollar industry.

Historically, the first practical examples of altering the concentration of ethylene within plants or plant products involved the combustion of organic materials such as wood or kerosene which unknowingly released ethylene as a constituent of the smoke or fumes. This technique was utilized to induce uniform flowering in pineapple (*Ananas comosus* L.)[1] and mango (*Mangifera indica* L.),[2] promote fruit ripening,[3] and citrus degreening.[4] It was also known that certain fruits (i.e., oranges) would release an unknown "emanation" that would ripen bananas (*Musa* spp.) when stored together onboard ship.[5] It was eventually established that the active agent in each case was ethylene[6,7] and that plants were themselves capable of synthesizing the hydrocarbon.[8] By 1946, a grower book describing the utilization of ethylene in tomato (*Lycopersicon esculentum* L.) ripening was published.[9]

As a gas, ethylene presents a unique set of requirements for its use in agriculture. Not only must the gas be in some way contained around the target tissue, but it is also highly explosive between 2.75 and 28.60% in air when ignited. Harvested products such as fruits do not present a major problem in that they can be placed in gas-tight rooms and thus exposed. Intact plants in the field, however, are another matter. Several attempts at field application of ethylene using portable tents or plastic ground covers were made for research purposes but seldom for commercial production. It was evident that a much more practical means of ethylene application was needed before the hormone could become a viable tool in the production phase of agriculture. This eventually led to the development of a number of nongaseous compounds that could be applied directly to the plant as an aqueous spray. These in turn either degraded to release ethylene on or in the plant, or stimulated the synthesis of ethylene by the tissue.

II. TYPES OF ETHYLENE-PRODUCING COMPOUNDS

Based on the site of origin of the ethylene molecule, ethylene-producing compounds can be separated into three general classes. The first class contains compounds which break down or are metabolized to release ethylene from the parent molecule (ethylene-releasing compounds). The second class of compounds induces the formation of ethylene by the target tissue (ethylene-inducing compounds) while the third releases ethylene held by an absorbant. Of the three classes, ethylene-releasing compounds have thus far proven to be the most consistently effective means of treating plants in the field with ethylene. Because of this, attention will primarily be focused on ethylene-releasing compounds in this chapter.

A. Ethylene-Releasing Compounds

A large number of ethylene-releasing compounds have been developed during the past 2 decades (Table 3). Many have utilized a chloroethyl group attached to either a phosphorus, silicon, or sulfur atom. Groups attached to other sites on the central atom are used to alter the stability of the chloroethyl group.

Table 1

ETHYLENE HAS BEEN SHOWN TO CAUSE OR BE IMPLICATED IN THE FOLLOWING PHYSIOLOGICAL RESPONSES IN SELECTED PLANTS AND PLANT PARTS

Stimulatory responses
    Adventitious root and root hair formation
    Root coiling
    Flowering in Bromeliaceae
    Abscission of leaves, flowers, and fruits
    Fruit dehiscence
    Fruit ripening
    Coleoptile growth
    Internode elongation
    Female flower production
    Root elongation
    Stomatal closure
    Germination of seeds, spores, and pollen
    Breaking dormancy in seeds, buds, tubers, corms, and bulbs
    Fruit growth
    Plumular hook closure
    Hypertrophy
    Exudation of fluids
    Tendril coiling
    Alteration of microfibril deposition
Inhibitory responses
    Cell elongation
    Internode elongation
    Photosynthesis
    Fruit growth
    Root growth
Involvement in
    Root and shoot geotropism
    Stress phenomenon — mechanical, wounding, water, nematode,
      fungi, and bacteria
    Senescence
    Apical dominance

Due to the large number of compounds, differences in their chemical properties, and the expense of developing a plant growth regulator for commercial use, only a few of these compounds have received significant attention. The primary ethylene-releasing compounds currently in commercial use are (2-chloroethyl)phosphonic acid (Ethrel®, Amchem 66-329, CEPA and ethephon), (2-chloroethyl)methylbis(phenylmethoxy)silane (Silaid®, or CGA-15281) and (2-chloroethyl)tris(2-methoxyethoxy)silane (Alsol®, CGA-13285, and etacelasil) (Figure 1). These specific compounds are believed to elicit physiological responses solely through the action of released ethylene, although the effect of other breakdown or metabolic products has been little explored. In plants capable of autocatalytic production of ethylene, ethylene released from the plant growth regulator accelerates the endogenous synthesis of ethylene within the tissue. This ethylene is thought, in several cases, to mediate the subsequent physiological response by the plant.

The first ethylene-releasing compound commercially developed was β-hydroxyethylhydrazine (Omaflora). β-Hydroxyethylhydrazine was initially found to have the ability to induce flowering in pineapple.[10] Its extremely low efficiency of conversion to ethylene ($\sim 1\%$), however, made it of only marginal value in eliciting ethylene-mediated responses.[11] Two other compounds were subsequently shown to release ethylene (monoethylsulfate and ethylpropylphosphate) but these also displayed low conversion efficiencies.[12,13] Both were utilized

Table 2
APPROVED USES FOR ETHREL IN
THE U.S.

Fruit ripening (postharvest)
   Banana, tomato (FL)
Fruit ripening
   Tomato, pepper
Fruit removal
   Apple, crabapple, carob, olive
Defoliation
   Cotton, roses, tallhedge buckthorn, apple
Fruit Loosening
   Apple, blackberry (WA, OR), cantaloupe, cherry (CA,
     AZ, TX), tangerine
Maturity and/or color development
   Apple, cranberry (MA, NJ, WI), fig (CA), filbert
     (OR), grape, pepper, pineapple, tomato (CA, TX.
     etc.)
Degreening (preharvest)
   Tangerine
Degreening (postharvest)
   Lemon
Dehiscence
   Walnut
Leaf curing
   Tobacco
Flower induction
   Pineapple, bromeliads
Sex expression
   Cucumber, squash
Flower bud development
   Apple
Lodging and plant height control
   Barley, daffodil, hyacinth, wheat
Stimulation of lateral branching
   Azalea, geranium

initially to some degree but were later dropped with the introduction of (2-chloroethyl)phosphonic acid.

The first major commercial ethylene-releasing compound, (2-chloroethyl)phosphonic acid was synthesized in 1946 by Kabachnik and Rossiiskaya[14] (Figure 1). It was not until 1963, however, that Maynard and Swan[15] described the ability of the molecule to release ethylene and proposed the decomposition mechanism involved. Other compounds, similar in general structure and breakdown, have been developed utilizing a sulfur moiety in the place of phosphonic acid.[16] The physiological activities of these compounds, in particular, (2-chloroethyl)hydroxymethylsulfone and (2-chloroethyl)sulfinic acid,[17] are in some ways comparable to (2-chloroethyl)phosphonic acid.

A group of (2-haloethyl)silanes developed by Ciba-Geigy has yielded several chemicals exhibiting promising characteristics. Two of these compounds have been in commercial use: (2-chloroethyl)methylbis(phenylmethoxy)silane and (2-chloroethyl)tris(2-methoxyethoxy)silane (Figure 1). These compounds have lower activation energy requirements for breakdown than (2-chloroethyl)phosphonic acid.

Several compounds must be metabolized by the plant tissue in order to release endogenous ethylene. These include intermediates in the synthesis pathway of ethylene itself, such as methionine, S-adenosylmethionine, and the immediate precursor to ethylene, 1-amino-

Table 3
ETHYLENE-RELEASING COMPOUNDS

(2-Chloroethyl)phosphonic acid

$$Cl-CH_2-CH_2-\overset{\overset{\textstyle O}{\|}}{\underset{\underset{\textstyle OH}{|}}{P}}-OH$$

β-(2-Chloroethyl)-2,2-dimethlyphosphonic dihydrazide

$$\underset{Cl-CH_2-CH_2}{\overset{CH_3}{\underset{|}{N}}}-NH-\overset{\overset{\textstyle O}{\|}}{\underset{\underset{\textstyle NH-NH_2}{|}}{P}}-H$$

(2-Chloroethyl)phosphonic acid dihydrazide

$$Cl-CH_2-CH_2-\overset{\overset{\textstyle O}{\|}}{\underset{\underset{\textstyle NH-NH_2}{|}}{P}}-NH-NH_2$$

Sulfones
  2-Chloroethyl-hydroxymethyl sulfone $Cl-CH_2-CH_2-SO_2-CH_2OH$
  2-Bromoethyl-hydroxymethyl sulfone $Br-CH_2-CH_2-SO_2-CH_2OH$
  2-Chloroethyl-1-hydroxyethyl sulfone $Cl-CH_2-CH_2-SO_2-COH-CH_3$
  2-Bromoethyl-1-hydroxyethyl sulfone $Br-CH_2-CH_2-SO_2-COH-CH_3$
  2-Chloroethyl-1-hydroxy-n-propyl sulfone $Cl-CH_2-CH_2-SO_2-CHOH-CH_2-CH_3$
  2-Bromoethyl-1-hydroxy-n-propyl sulfone $Br-CH_2-CH_2-SO_2-CHOH-CH_2-CH_3$
  2-Chloroethyl-1-hydroxy-n-butyl sulfone $Cl-CH_2-CH_2-SO_2-CHOH-CH_2-CH_2-CH_3$
  2-Bromoethyl-1-hydroxy-n-butyl sulfone $Br-CH_2-CH_2-SO_2-CHOH-CH_2-CH_2-CH_3$

2-Chloroethyl-1-hydroxy-iso-butyl sulfone

$$Cl-CH_2-CH_2-SO_2-\overset{\overset{\textstyle CH_3}{|}}{C}OH-CH_2-CH_3$$

2-Bromoethyl-1-hydroxy-iso-butyl sulfone

$$Br-CH_2-CH_2-SO_2-\overset{\overset{\textstyle CH_3}{|}}{C}OH-CH_2-CH_3$$

2-Chloroethyl sulfinic acid

$$Cl-CH_2-CH_2-\overset{\overset{\textstyle O}{\|}}{S}-OH$$

Silanes
  (2-Chloroethyl)methylbis(phenylmethoxy)silane

$$Cl-CH_2-CH_2-\overset{\underset{\underset{\textstyle CH_3}{|}}{}}{Si}(O-CH_2-C_6H_5)_2$$

  (2-Chloroethyl)tris(2-methoxyethoxy)silane $Cl-CH_2-CH_2-Si(O-CH_2-CH_2-OCH_3)_3$
  (2-Chloroethyl)tris(methoxy)silane $Cl-CH_2-CH_2-Si(O-CH_3)_3$
  (2-Chloroethyl)tris(isopropoxy)silane $Cl-CH_2-CH_2-Si[O-CH(CH_3)_2]_3$
  (2-Chloroethyl)tris(lauroxy)silane $Cl-CH_2-CH_2-Si(O-C_{12}H_{25})_3$

2-Trimethoxysilylethyl ethyl methsulfonium iodide

$$(CH_3O)_3Si-CH_2-CH_2-\overset{\overset{\textstyle I^-}{}}{\underset{\underset{\textstyle CH_3}{|}}{S^+}}-CH_2-CH_3$$

Poly-2-chloroethylmethoxysilane $(Cl-CH_2-CH_2)_{3,2}Si(O-CH_3)_{1,2}$
2-Chloroethylsilane $Cl-CH_2-CH_2-SiH_3$

(2-Chloroethyl)acetoxybis(methoxy)silane

$$Cl-CH_2-CH_2-\overset{\overset{\textstyle O-COCH_3}{|}}{Si}(O-CH_3)_2$$

Table 3 (continued)
ETHYLENE-RELEASING COMPOUNDS

(2-Chloroethyl)ethylbis(methoxy)silane

$$\text{Cl—CH}_2\text{—CH}_2\text{—Si(O—CH}_3)_2 \quad \overset{\displaystyle \text{CH}_2\text{—CH}_3}{|}$$

(2-Chloroethyl)bis(methoxy)chlorosilane

$$\text{Cl—CH}_2\text{—CH}_2\text{—Si(O—CH}_3)_2$$
$$|$$
$$\text{Cl}$$

(2-Bromoethyl)tris(phenyl)silane $\text{Br—CH}_2\text{—CH}_2\text{—Si(C}_6\text{H}_5)_3$

Biphenyl-2(trimethylsilyl)ethyl phosphate $\text{(CH}_3)_3\text{Si—CH}_2\text{—C(C}_6\text{H}_5)_2\text{—PO}_3\text{H}_2$

(2-Chloroethyl)butoxybis(methyl)silane $\text{Cl—CH}_2\text{—CH}_2\text{—Si(CH}_3)_2$
$$|$$
$$\text{O—CH}_2\text{(CH}_2)_2\text{—CH}_3$$

(2-Chloroethyl)tris(methyl)silane $\text{Cl—CH}_2\text{—CH}_2\text{—Si(CH}_3)_3$
(2-Chloroethyl)tris(ethoxy)silane $\text{Cl—CH}_2\text{—CH}_2\text{—Si(O—CH}_2\text{—CH}_3)_3$
(2-Chloroethyl)tris(acetoxy)silane $\text{Cl—CH}_2\text{—CH}_2\text{—Si(O—CO—CH}_3)_3$
(2-Chloroethyl)tris(propanoxy)silane $\text{Cl—CH}_2\text{—CH}_2\text{—Si(O—CH}_2\text{—CH}_2\text{—CH}_3)_3$
(2-Chloroethyl)tris(caprooxy)silane $\text{Cl—CH}_2\text{—CH}_2\text{—Si(O—CO—CH}_2\text{(CH}_2)_3\text{CH}_3)_3$
(2-Chloroethyl)acetoxybis(ethoxy)silane $\text{Cl—CH}_2\text{—CH}_2\text{—Si(O—CH}_2\text{—CH}_3)_2$
$$|$$
$$\text{O—CO—CH}_3$$

(2-Chloroethyl)tris(butoxy)silane $\text{Cl—CH}_2\text{—CH}_2\text{—Si(O—CH}_2\text{(CH}_2)_2\text{CH}_3)_3$
(2-Chloroethyl)trichlorosilane $\text{Cl—CH}_2\text{—CH}_2\text{—SiCl}_3$
(2-Chloroethyl)methyldichlorosilane $\text{Cl—CH}_2\text{—CH}_2\text{—SiCl}_2$
$$|$$
$$\text{CH}_3$$

(2-Chloroethyl)bis(ethoxy)chlorosilane $\text{Cl—CH}_2\text{—CH}_2\text{—Si(O—CH}_2\text{—CH}_3)_2$
$$|$$
$$\text{Cl}$$

Metabolized ethylene-releasing compounds

1-Aminocyclopropane-1-carboxylic acid

$$\begin{array}{c} \text{CH}_2\text{——CH}_2 \\ \diagdown\text{C}\diagup \\ \text{NH}_2\diagup \ \diagdown\text{COOH} \end{array}$$

N-formyl-1-aminocyclopropane-1-carboxylic acid

$$\begin{array}{c} \text{CH}_2\text{——CH}_2 \\ \diagdown\text{C}\diagup \\ \text{COH—NH}\diagup \ \diagdown\text{COOH} \end{array}$$

2-Amino-4-(methylthio)butanoic acid

$$\text{CH}_3\text{—S—CH}_2\text{—CH}_2\text{—CH—COOH} \quad \overset{\displaystyle \text{NH}_2}{|}$$

Early ethylene-releasing compounds
β-Hydroxylethyl hydrazine $\text{HO—CH}_2\text{—CH}_2\text{—NH—NH}_2$
Ethyl hydrazine $\text{CH}_3\text{—CH}_2\text{—NH—NH}_2$
Sym-diethyl hydrazine $\text{CH}_3\text{—CH}_2\text{—NH—NH—CH}_2\text{—CH}_3$

Unsym-bis-(2-hydroxylethyl)hydrazine

$$\begin{array}{c} \text{HO—CH}_2\text{—CH}_2 \\ \diagdown\text{N—NH}_2 \\ \text{HO—CH}_2\text{—CH}_2\diagup \end{array}$$

Aminomorpholine

$$\text{O}\diagdown\begin{array}{c}\text{H}_2\text{C—CH}_2\\ \\ \text{H}_2\text{C—CH}_2\end{array}\diagdown\text{N—NH}_2$$

Table 3 (continued)
ETHYLENE-RELEASING COMPOUNDS

2-Hydroxyl-N-(2-hydroxylethyl)carbazinate $HO-CH_2-CH_2-NH-\bar{N}H-CO-O-CH_2-CH_2OH$

2-(2-Hydroxyethyl)semicarbazine $HO-CH_2-CH_2-N-NH_2$
$$|$$
$$CO-NH_2$$

Ethylpropylphosphonate $CH_3-CH_2-POOH-O-CH_2-CH_2-CH_3$
Monoethylsulfate $CH_3-CH_2-O-SO_3H_2$
Monoethylphosphate $CH_3-CH_2-O-PO_3H_2$
Diethyl phosphate $CH_3-CH_2-O-PO_2-O-CH_2-CH_3$
Butyric acid $CH_3-CH_2-CH_2-COOH$
Ethyl iodide CH_3-CH_2I
Ethanol CH_3-CH_2OH
3-(Methylthio)propionaldehyde $CH_3-S-CH_2-CH_2-COH$

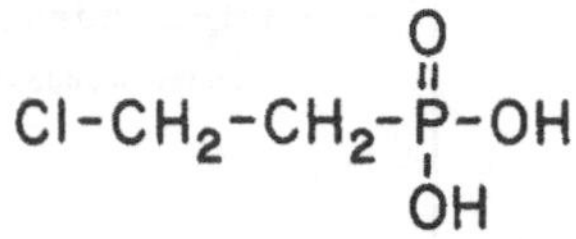

$$Cl-CH_2-CH_2-\overset{\overset{\displaystyle O}{\|}}{\underset{\underset{\displaystyle OH}{|}}{P}}-OH$$

(2-chloroethyl)phosphonic acid

Ethrel

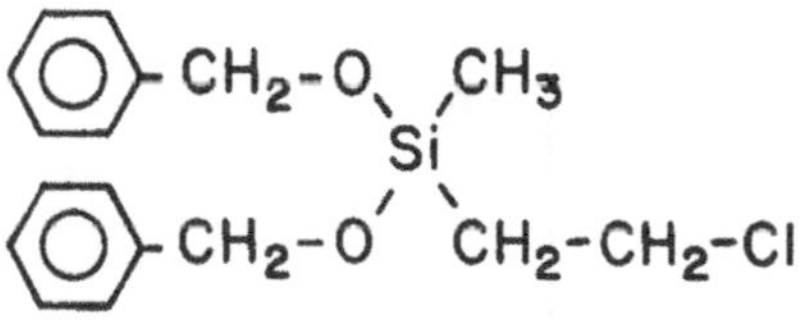

(2-chloroethyl)methylbis(phenylmethoxy)silane

Silaid

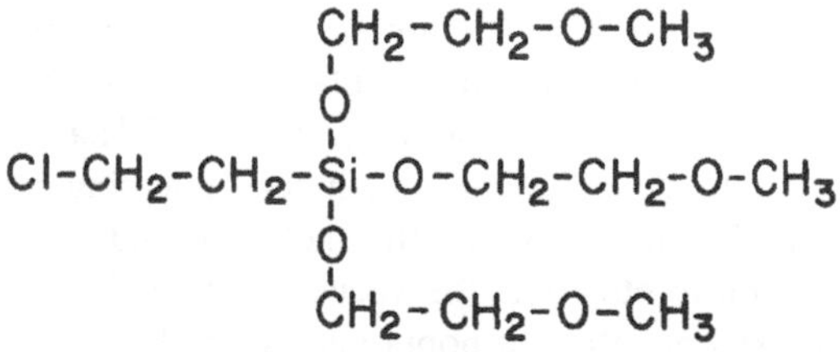

$$Cl-CH_2-CH_2-\underset{\underset{\displaystyle CH_2-CH_2-O-CH_3}{\underset{\displaystyle |}{O}}}{\overset{\overset{\displaystyle CH_2-CH_2-O-CH_3}{\overset{\displaystyle |}{O}}}{Si}}-O-CH_2-CH_2-O-CH_3$$

(2-chloroethyl)tris(2-methoxyethoxy)silane

Alsol

FIGURE 1. Structures of the most widely used ethylene-releasing compounds in agriculture.

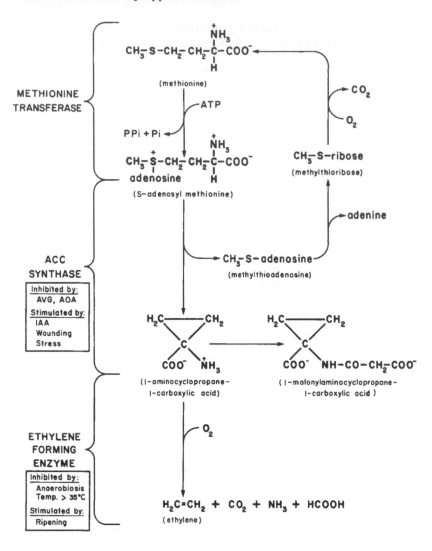

FIGURE 2.   Biological route for ethylene synthesis indicating points of pathway regulation.

cyclopropane-1-carboxylic acid (ACC)[18] (Figure 2). Due to its proximity to the final product and its apparently relatively specific role in the plant, ACC has thus far proven to be the most effective and commonly used compound of this type. ACC has not, however, displayed consistently satisfactory results when used as an ethylene-releasing compound. This appears to be due in part to variation in uptake of the molecule and movement to the primary subcellular site of conversion to ethylene, the vacuole.[19] In addition, ACC is readily translocated acropetally in the xylem, often to nontarget tissue.[20-22]

B. Ethylene-Inducing Compounds

A number of chemically diverse compounds are capable of inducing the formation of ethylene by plants. These may be grouped, however, into two general classes based on their mode of action. Chemicals in one class stimulate the synthesis of ethylene through a direct or relatively direct effect on the synthesis pathway. Most affect ACC synthase which appears under normal conditions to be the rate-limiting step in ethylene synthesis. Auxins and auxin analogs such as indoleacetic acid (IAA), α-naphthaleneacetic acid (NAA), 2,4-dichlorophenoxyacetic acid (2,4-D), and others are known to act in such a manner.[23-25] Application of auxins to plant tissues result in typical auxin-like responses as well as those attributable to ethylene (e.g, leaf epinasty, chlorophyll destruction, sex expression, fruit ripening, root hair initiation, and others).[26] The complexity of the interaction between these two hormones often precludes use of auxins for the induction of ethylene-mediated responses. This is especially true for physiological responses in which there is an antagonistic effect between the two hormones (e.g., abscission, dehiscence, and apical dominance).[26,27]

Compounds grouped into the second class act indirectly by wounding or stressing the tissue which subsequently accelerates the ethylene synthesis pathway. The enhanced production of ethylene from wounded or stressed plants is a well-documented phenomenon.[28-30] Cooper and Henry surveyed a large number of potential ethylene-inducing compounds.[31,32] They found that the effective compounds appeared to act through the induction of wound or stress ethylene. Compounds such as ascorbic acid, citric acid, cycloheximide, and iodoacetic acid injure the rind of citrus fruits stimulating the production of ethylene which weakens the fruit abscission zone.

C. Adsorbed Ethylene Compounds

Compounds which contain adsorbed ethylene, such as Ethad®, also show considerable promise.[33] Molecular sieves of crystalline aluminosilicates, activated by the removal of water, adsorb ethylene molecules into uniform microscopic pores.[34] Other adsorbants such as activated charcoal, kieselguhr, bentonite, Fuller's earth, brick dust, and silica gel do not exhibit the same degree of effectiveness as molecular sieves. The rate of ethylene release is controlled by mixing the powdered adsorbent with additives (e.g., oils) which act as barriers to ethylene diffusion. The relative efficiency of ethylene retention and subsequent release can potentially be altered by varying the sieve content of the mixture and type of additive used.

III. PLANT RESPONSES TO ETHYLENE-RELEASING COMPOUNDS

The use of ethylene-releasing compounds has become extremely important in the production phase of agriculture. These compounds are utilized to induce responses as diverse as abscission, growth inhibition, flowering, ripening, maturation, freeze and disease resistance, and the stimulation of latex flow. Application usually involves aqueous sprays of the active compound which is either dissolved in water, [e.g., (2-chloroethyl)phosphonic acid] or suspended in an emulsion, [e.g, (2-chloroethyl)silanes].

A. Abscission

The abscission response of a variety of species to ethylene has been well documented.[35] In many cropping situations, the promotion of abscission can be highly beneficial. Ethylene-releasing compounds are currently used to promote defoliation, thinning of blossoms and young fruit, and the loosening, abscission, or dehiscence of mature fruit.

Defoliation of cotton (*Gossypium hirsutum* L.) improves the efficiency of mechanized harvest and reduces cotton-fiber staining.[36] This is accomplished with (2-chloroethyl)phosphonic acid (8000 ppm) which gives approximately 75% defoliation by the 8th day after application. Higher temperatures increase the effectiveness of the treatment.[37] Removal of diseased foliage on snap beans with 1000 ppm of (2-chloroethyl)phosphonic acid applied 3 to 5 days before harvest has also been shown to aid mechanical harvest.[38]

Early defoliation of deciduous ornamentals and fruit trees allows an earlier start to harvesting.[39] This lengthens the harvest period and permits more efficient utilization of storage, machinery, and labor. Multiple sprays ranging from 250 to 2500 ppm of (2-chloroethyl)phosphonic acid appear to give better results than a single, concentrated spray.[37] Additives such as urea, potassium iodide, and bromodine are used for some species to enhance the defoliation response.[38] In fact, potassium iodide and bromidine by themselves mimic some ethylene effects such as abscission.[40,41] Some species, such as peach and pear, incur excessive damage from these treatments. Serious twig dieback and bud death have been reported with single applications of higher concentrations (2000 to 4000 ppm).[40]

Ethylene-releasing compounds are increasingly being used to thin blossoms and/or immature fruits of several crop species to allow adjustment of fruit load. This results in fewer but larger fruits of higher quality at harvest. Use of ethylene-releasing compounds in place of hand labor reduces labor costs and minimizes the time needed for completion of the operation.

Foliar sprays of (2-chloroethyl)phosphonic acid (250 to 500 ppm) when applied at 20 to 100% full bloom have been effective for thinning peaches (*Prunus persica* [L.] Batsch.).[42,43] Thinning at the blossom stage, however, increases the risk of insufficient fruit load due to subsequent pollination deficiencies or late frost. If (2-chloroethyl)phosphonic acid is applied at a later developmental stage, it must be done during the 4 or 5 days of fruit cytokinesis or the concentration required for thinning can cause excessive leaf abscission.[43-46] In addition to leaf loss, excessive (2-chloroethyl)phosphonic acid has also been shown to cause gummosis, terminal dieback, and enlarged lenticels.

Sprays of (2-chloroethyl)methylbis(phenylmethoxy)silane have been effective in thinning peaches when applied at full bloom (1500 to 2750 ppm).[47,48] Unlike (2-chloroethyl)phosphonic acid, postbloom sprays of 125 to 480 ppm of this compound, when the ovule is approximately 12 mm long, give excellent thinning without undesirable side effects.[49,50] The sensitivity of the young fruits appears to increase with increasing ovule length.[49]

Several other crops have been thinned at bloom or postbloom using (2-chloroethyl)phosphonic acid, e.g., grapes (*Vitis vinifera* L., 10 to 1000 ppm), coffee (*Coffea arabica* L., 400 to 600 ppm), pears (*Pyrus communis* L., 240 to 480 ppm), and prunes (*Prunus domestica* L., 50 to 150 ppm).[37,51,52] The higher concentrations used for grape and pear thinning resulted in complete removal of blossoms and immature fruit. With prunes, applications for three consecutive years resulted in no apparent damage to vigorous trees, however, weak or stressed trees developed excessive gummosis.[52]

The use of (2-chloroethyl)phosphonic acid for thinning apples (*Malus sylvestris* L.) has generally been unsuccessful. Concentrations of 1000 to 2000 ppm applied prebloom and early postbloom over thinned. Applications at later developmental stages tended to thin adequately, but produced unacceptable phytotoxic effects.[38]

Ethylene-releasing compounds have been used to loosen fruit by promoting the partial formation of the abscission zone, thereby reducing the force needed to remove fruit at harvest. This is especially beneficial for mechanically harvested crops with slender branches which poorly transmit the force imparted by the harvesting machinery. With mechanical harvesting, reduced removal force decreases the shaking time, increases the percentage of fruit removed, and minimizes damage to the fruit and plant.[53]

Sweet (*Prunus avium* L.) and sour (*P. cerasus* L.) cherries are effectively loosened with 250 to 300 ppm of (2-chloroethyl)phosphonic acid applied 7 to 14 days prior to harvest.[53-55] Improved quality of the harvested fruit and decreased incidence of persisting pedicels were noted in some cases.[53,56] Rates above 1000 ppm are generally phytotoxic. Both (2-chloroethyl)methylbis(phenylmethoxy)silane   and   (2-chloroethyl)tris(2-methoxyethoxy)silane have also been tested as cherry fruit harvest aids, the former proving more effective.[57]

Generally, higher rates (e.g., 500 to 2000 ppm) of (2-chloroethyl)phosphonic acid are needed to decrease the fruit removal force of apples.[58] This treatment tends to counteract the delay in abscission caused by the use of succinic acid 2,2-dimethyl hydrazide (Alar®) for enhanced fruit color.[58]

Loosening of citrus fruit for harvest has been attempted using ethylene-releasing compounds, but with little success. Often poor loosening and unacceptable defoliation and chlorosis occurred with these treatments.[31,59] Better results have been obtained with ethylene-inducing compounds applied singly or in combinations.[31,32,60,61] Chemicals such as cyclo-heximide (Acti-Aid®) (2 to 25 ppm), glyoxal dioxime (Pik-off®) (75 to 150 ppm), 5-chloro-3 methyl-4-nitro-1H pyrazole (Release®) (75 to 125 ppm) and chlorothalonil (Sweep®) (100 to 200 ppm) have been used with relatively good success.

Olive (*Olea europaea* L.) fruit may be loosened using (2-chloroethyl)phosphonic acid,[62-65] however, unacceptable leaf loss generally occurs.[62-64] Other compounds, most notably the (2-haloethyl)silanes, show more promise. Both (2-chloroethyl)methylbis(phenyl-methoxy)silane and (2-chloroethyl)tris(2-methoxyethoxy)silane loosen olive fruit without unacceptable levels of leaf drop.[62,64,66,68]

Additives such as antitranspirants[69] and urea[27] enhance olive fruit loosening and permit lower levels of ethylene-releasing compounds to be used. Antitranspirants,[69] naphthalene acetic acid,[27] and calcium sprays[70] have been shown to minimize or decrease leaf abscission accompanying fruit loosening.

The dehiscence of nut crops such as almonds (*Prunus amygdalus* Batsch.), walnuts (*Juglans regia* L.), pecans (*Carya illinoenis* [Wang] K. Koch.), Chinese chestnuts (*Castanea mollissima* Blume), and filberts (*Corylus avellana* L.) can be initiated with ethylene-releasing compounds to facilitate harvest.[38] However, in some cases (e.g., pecans and Chinese chestnuts), the concentration required to induce dehiscence also results in considerable leaf abscission.[71,72] This is generally detrimental to yield in the subsequent growing season.

B. Growth Effects

Ethylene has long been known to modulate a number of growth responses in plants (e.g., shoot elongation, apical dominance, root initiation, and others).[35,73] Many of these growth responses can be beneficially altered in some species using ethylene-releasing compounds.

Lodging of small grains reduces crop yield. The use of (2-chloroethyl)phosphonic acid tends to shorten and stiffen grain bearing stems and results in less lodging. Applications of 0.28 to 1.12 kg/ha of (2-chloroethyl)phosphonic acid after tillering are effective, often increasing harvestable yields. Early application of 0.56 to 2.24 kg/ha also enhances tillering[38] however, this does not always increase yield since grain heads are often smaller. Soybeans (*Glycine max* Merr.) may similarly benefit from sprays of 0.56 kg/ha or less.[38]

Reduction of terminal growth of several woody plant species to control plant size and increase flower bud formation can be accomplished with (2-chloroethyl)phosphonic acid. For example, sprays of 250 to 1000 ppm on apples decreases terminal growth and increases the number of flower buds. Lateral bud break is also enhanced.[74] Similar results have been obtained with a number of ornamentals (e.g., azalea, cotoneaster, hibiscus, hydrangea, petunia, privet, poinsettia, rose, and pine).[38] In grapes, excessive vigor of Cabernet Sauvignon vines is most effectively reduced by 500 ppm foliar sprays although the effect is highly temperature dependent.[75]

C. Flowering

Ethylene-releasing compounds are used to induce flowering or promote the formation of pistillate flowers in several species. In bromeliads such as the pineapple, flowering can be synchronized using ethylene-releasing compounds. This decreases the number of times the crop must be harvested.[76] Flowering of ornamental bromeliads may be similarly induced.[38]

Increased flower production has been noted for apples[77,78] and pears due to fall or spring applications of (2-chloroethyl)phosphonic acid. Fall applications on sweet cherry and plum (*Prunus domestica* L.) result in a delay in flowering in the spring[79] which could be advantageous in areas with late spring frosts. Effective concentrations, however, generally exhibit some phytotoxicity (i.e., gummosis, bud abscission or failure to open, and reduced fruit set) which minimizes their potential usefulness.

Ethylene-releasing compounds have been shown to significantly alter the pattern of sex expression in many of the cucurbits.[80-82] During early growth, the plant produces almost exclusively staminate flowers. Pistillate flowers are normally not produced until the plant has achieved some size. Application of an ethylene-releasing compound at the first or second true leaf stage, however, results in a much earlier induction of pistillate flowers.[80,81] In some cucumber (*Cucumis sativa* L.) and squash (*Cucurbita pepo* L.) cultivars, this results in an earlier first harvest and higher total yields. In andromonecious lines, it may be used for seed production.[80,82]

D. Maturation and Ripening

The involvement of ethylene in fruit maturation and ripening has been of interest throughout much of this century.[83] Exogenous ethylene initiates the ripening response in climacteric fruits. It will also accelerate desirable changes in some nonclimacteric fruits (e.g., chlorophyll loss in citrus).[4]

The induction and acceleration of normal ripening is of particular value in crops that are to be gathered in a single harvest. For example, the application of (2-chloroethyl)phosphonic acid to tomatoes results in a higher percentage of ripe fruit at harvest.[84] Ethylene-releasing compounds have also been shown to advance the maturity of apple, pear, peach, prune, fig (*Ficus carica* L.), and cranberry (*Vaccinium macrocarpon* Ait.) fruit.[45,51,85-88]

Citrus fruits are degreened and banana fruits ripened by exposure of the harvested fruit to ethylene. In areas where ripening rooms are not available, ethylene-releasing compounds have been shown to give similar results. Bananas treated with 500 to 2000 ppm of (2-chloroethyl)phosphonic acid ripen normally with no apparent quality loss.[89] Citrus, dipped in a 1000 ppm solution, degreened satisfactorily in 7 to 10 days. Preharvest foliar sprays on tangerines (*Citrus reticulata* Blanco) decreased the time required for degreening and reduced the incidence of decay during storage.[37]

Some crops harvested for vegetative plant parts, such as sugarcane (*Saccharum officinarum* L.) and tobacco (*Nicotiana tabacum* L.), also benefit from sprays of (2-chloro-ethyl)phosphonic acid. The accumulation of sugar in the stems of sugarcane can be promoted by 2.2 to 4.4 kg/ha of (2-chloroethyl)phosphonic acid which promotes early harvest and/or increased yields.[37,90] Maturation of tobacco is characterized by leaf chlorophyll loss and certain other biochemical changes. Application of (2-chloroethyl)phosphonic acid (1.1 to 2.2 kg/ha) accelerates maturation and reduces curing time. This decreases losses due to sunburn, barn rot, and other curing problems.[91] The same ethylene-releasing compound has been shown to inhibit sucker growth of topped plants and reduce nicotine content of the leaves.[38,91]

E. Latex Production

Obtaining latex from *Hevea brasiliensis* Muell. for rubber products is hindered by latex coagulation which plugs the latex vessels. Ethylene-releasing compounds decrease the rate at which plugging occurs and greatly increase latex yield.[38,92,93] These are generally applied in an oil carrier banded below the site of tapping. Tapping is normally done as a half spiral every 2 days; however, the use of an ethylene-releasing compound allows less frequent tapping, and/or shorter incisions. This potentially reduces labor requirements and lengthens the economic life of the tree.[34] Both (2-chloroethyl)phosphonic acid[33] and (2-halo-ethyl)silanes[93] have proven effective in improving yield. The (2-haloethyl)silanes offer the additional benefit of being less subject to leaching by rainfall due to their low water solubility.[33,34]

F. Other Applications

Ethylene-releasing compounds have also been shown to have potential for increasing disease resistance, frost tolerance, and bud hardiness in selected species. In sweet potato (*Ipomoea batatas* L. Lam), disease resistance of storage roots was improved after exposure to exogenously applied ethylene as well as ethylene induced by infection with pathogenic and nonpathogenic strains of the fungus *Ceratocystic fimbriata* Ell. and Halst.[94] Chocolate spot, a physiological disease of the Irish potato (*Solanum tuberosum* L.) was controlled by four sprays of 200 ppm (2-chloroethyl)phosphonic acid at 2-week intervals.[95]

Frost tolerance of tomato was improved by treating the transplants with 3000 ppm sprays 12 days prior to pulling for transplanting.[96] The treatment tended to harden plants to frost injury as well as induce stem thickening which was also associated with greater frost tolerance. Reduced bud injury of stone fruits to freezing temperatures with sprays of (2-chloroethyl)phosphonic acid appears to act through two modes. Summer and fall sprays of 200 to 2000 ppm provided a gain of 1 to 4°C freeze protection for buds of apricots (*Prunus araeniaca* L.), cherries, prunes, and peaches.[97] Treatment (30 to 200 ppm) also delayed the time of bloom in the spring which resulted in reduced frost injury of sweet cherry and plum.[79] In both cases, however, the benefits derived from improved protection were offset by damage incurred from the treatment.

IV. FACTORS INFLUENCING THE EFFECTIVENESS OF ETHYLENE-RELEASING COMPOUNDS

Characteristically, there is an innate degree of variability in the magnitude of the physiological response to most exogenously applied nonlethal plant growth regulators. This is especially true of the response to ethylene-releasing compounds where the actual chemical applied to the plant does not directly elicit the response, but rather acts primarily as a means of delivery of the active molecule. The release of ethylene and the subsequent response by the plant are affected by a number of chemical and plant parameters and their interaction with the environment. The complexity of this relationship with ethylene-releasing compounds increases the potential for variation.

The response of plants or plant parts to ethylene can be grouped into three general classes based on the width of the concentration range of applied chemical giving the desired physiological response and the severity of the results obtained when this range is exceeded.[139]

1. Critical physiological responses are those wherein the treated plants incur serious damage when the amount of the applied chemical exceeds the range necessary to elicit a desired response. Damage can result from an excessively high level of ethylene released or prolonged exposure to lower levels. Representative examples of this group would include peach fruit thinning[44] and pecan shuck dehiscence.[72] Excess ethylene can result in immediate crop loss in the peach through removal of all of the young fruits or a delayed effect as in the pecan where the next season yield is drastically reduced with premature defoliation of the tree.
2. The intermediate class of ethylene responses are those in which damage is sustained from excessive ethylene, but the losses are much more limited in relation to the previous class. Examples of this would be the ripening of pepper (*Capsicum* spp.) fruit on the plant[98] or flower abscission of ornamental olive.[99] Excess ethylene results in damage, however, generally not in total crop loss (peppers) or permanent damage to the plant (ornamental olive).
3. Noncritical responses are those in which there is a substantial safety margin in the concentration that may be applied without significant damage to the crop. Examples would be tomato ripening,[84] enhanced latex flow in rubber plant,[100] and leaf abscission in a number of ornamentals.[39]

While a classification of this nature is somewhat arbitrary due to the subjective nature of the "extent of injury" factor and the differential sensitivity of certain tissues with age and condition, it allows focusing on an important element of the use of ethylene-releasing compounds. Responses which are less critical in their safety margins tend to be among the first to be integrated into common production practices. As we proceed toward resolving the requirements of more critical ethylene-mediated responses, the demand for a complete understanding of the operative forces controlling the response also increases. As a consequence, a relatively complete understanding of the chemical and plant factors and their various interactions is essential.

A. Chemical Factors

The intrinsic chemical characteristics of each ethylene-releasing compound strongly influence its field response and, therefore, its subsequent utility in agriculture. Characteristics of primary importance are those which significantly affect the release of ethylene from the parent molecule such as mechanism of breakdown, energy of activation, solubility, and pH stability. In addition, the occurrence of side reactions during breakdown and the formation of undersirable breakdown products or metabolites are also of considerable interest.

1. Chemistry of the Molecule

For many ethylene-releasing compounds, the relationship between changing environmental conditions (temperature, light, relative humidity, wind, and precipitation) and the basic chemical properties of the molecule, relative to the release of ethylene, has been little explored. It is of little surprise then, to frequently find significant differences in optimum concentration for a given crop occurring between different locations and times of application.

The mode of breakdown of the parent molecule and the subsequent release of ethylene has important implications for the effectiveness of the ethylene-releasing compound.[140-142] For many of the 2-haloethyl compounds, initial breakdown is believed to occur upon hydrolysis of the chloroethyl bond in an aqueous solution.[17] This may occur on the surface of the plant or within the cytoplasm. For (2-chloroethyl)phosphonic acid, a nucleophilic attack by water or a hydroxyl ion on the central phosphorus atom is believed to lead to the simultaneous elimination of chlorine and liberation of ethylene from the parent molecule.[15,101]

The stability of most ethylene-releasing compounds is markedly altered by pH.[11,15,102,103] At low pH, (2-chloroethyl)phosphonic acid is relatively stable, however, above pH 4.5 an increasing percentage of the molecules is converted to the dianionic form thereby becoming susceptible to hydrolysis.[15,102] Decomposition follows first order kinetics in relation to the dianionic fraction which approaches unity at pH 9 and above.[103] Thus, solution pH can also strongly influence the rate of release of ethylene from the parent molecule. Increasing the pH of the aqueous solution or the penetration of the molecule into the cytoplasm, where the pH is more conducive for rapid degradation of the compound, can greatly affect release kinetics.

The energy of activation also significantly affects the release of ethylene. The greater the activation energy of a molecule, the higher the energy requirements for its uncatalyzed breakdown. Consequently, ethylene-releasing compounds with high activation energies are more temperature dependent than those with low activation energies (Figure 3).[143] When the field temperature is cool, much less ethylene is released from compounds with high activation energies, thus temperature is more critical. The effect of temperature on the release of ethylene from (2-chloroethyl)methylbis(phenylmethoxy)silane spray on Teflon® is illustrated in Figure 4.[143] The point at which 50% of the parent compound has degraded and released ethylene occurs at 50, 132, and 786 min for 35, 26 and 10°C, respectively. Temperature range and variability during treatment may then be important considerations in both the selection of an appropriate ethylene-releasing compound and the concentration applied.

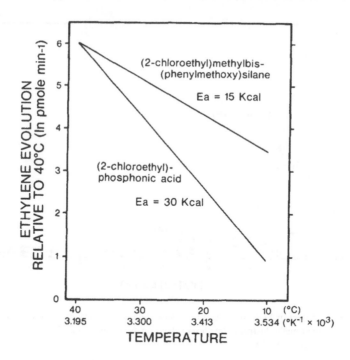

FIGURE 3. Effect of temperature on the release of ethylene from (2-chloroethyl)methylbis(phenylmethoxy)silane and (2-chloroethyl)phosphonic acid. Activation energies are 15 and 30 Kcal, respectively.

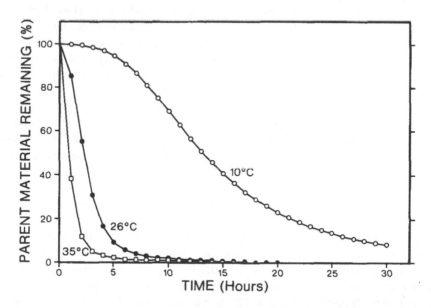

FIGURE 4. Effect of temperature on the loss of (2-chloroethyl)methylbis(phenylmethoxy) silane at 10, 26, and 35°C as determined by ethylene evolution.

range and variability during treatment may then be important considerations in both the selection of an appropriate ethylene-releasing compound and the concentration applied.

Atmospheric relative humidity at the time of application can strongly modulate the ethylene-release kinetics of some compounds. Up to a point, increasing water vapor pressure facilitates the breakdown of (2-chloroethyl)phosphonic acid.[104] The optimum vapor pressure

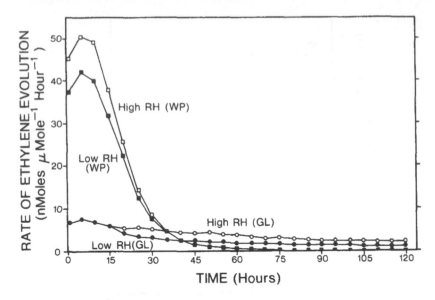

FIGURE 5.   Effect of high (62%) and low (36%) relative humdity on the release kinetics of (2-chloroethyl)methylbis(phenylmethoxy)silane on wax paper and (2-chloroethyl)phosphonic acid on glass at 25°C.

value increases with temperature, doubling for each 10°C increase. Water vapor is thought to hydrolyze the compound after it has dried to a film.

High relative humidity following application tends to decrease the rate of drying of the spray droplets,[105] which prolongs the period of aqueous hydrolysis. Thus, higher relative humidity can result in both higher rates of ethylene release and a greater total yield (Figure 5). While (2-chloroethyl)phosphonic acid maintains an elevated rate of ethylene evolution for an extended time, higher rates are found only initially with (2-chloroethyl) methylbis(phenylmethoxy)silane, apparently due to the more rapid disappearance of the parent molecule.

Light may also affect the rate of breakdown of some ethylene-releasing compounds. In experiments where the reaction temperature was carefully controlled, light altered the rate of ethylene release from (2-chloroethyl)methylbis(phenylmethoxy)silane. The response was affected by the surface characteristics of the material upon which the parent compound was sprayed and could not be explained by slight changes in temperature. High light (620 μmol m^{-2} sec^{-1}) accelerated the initial rate of breakdown when the chemical was sprayed as an aqueous solution on wax paper or aluminum foil. High light also resulted in reduced yield after 120 hr at which time the rate of breakdown was essentially zero (Figure 6). However, light had no effect on yield from application to glass. Interactions between the surface of the target tissue and the ethylene-releasing compound alter the rate of ethylene evolution and subsequent yield.[143] The rate of ethylene production from (2-chloroethyl) methylbis(phenylmethoxy)silane was strongly retarded when the parent molecule was in contact with glass (Figure 7). Yield at 120 hr was also reduced. The reduced breakdown is evidently due to the formation of side reactions such as those found in the soil (also predominantly silicon dioxide), which yield derivatives with increased stability.[106] Wax paper, aluminum foil, and Teflon® surfaces permitted progressively greater rates of decomposition, probably reflecting successively lower degrees of interaction.

The water solubility of the parent molecule may also significantly influence its response in the field. More hydrophilic ethylene-releasing compounds appear to exhibit greater potential movement once within the plant. Most breakdown, however, appears to occur on the

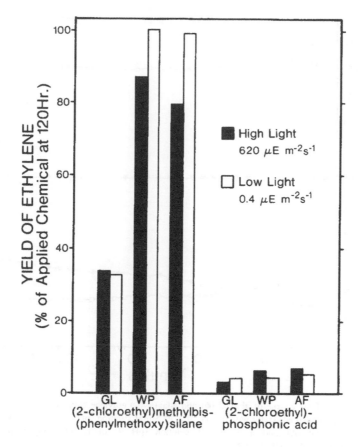

FIGURE 6.   Effect of high (620 μE m^{-2} sec^{-1}) and low (0.4 μmol m^{-2} sec^{-1}) light intensity on the yield of ethylene after 120 hr from (2-chloroethyl)methylbis(phenylmethoxy)silane and (2-chloroethyl)phosphonic acid on glass, wax paper, and aluminum foil, 25°C.

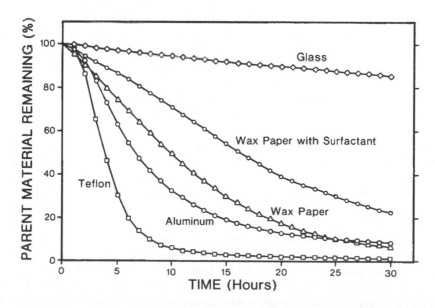

FIGURE 7.   Effect of target surface chemistry on the release kinetics of ethylene from (2-chloroethyl)methylbis(phenylmethoxy)silane at 25°C.

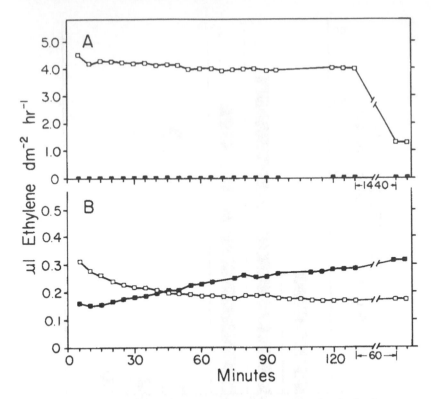

FIGURE 8.   Time course of abaxial (■) and adaxial (□) ethylene flux for hemistomatous peach leaves treated adaxially with 10 μℓ of a 19.5 mM solution (pH 7) of (2-chloroethyl) phosphonic acid (A) and a 6.75 mM solution (pH 7) of (2-chloroethyl)methylbis (phenylmethoxy)silane (B) in light.

surface of the target organ rather than after penetration. An exception to this would be 1-aminocyclopropane-1-carboxylic acid which produces ethylene through the metabolism of the parent molecule by the plant.[107] Molecules with greater lipid solubility tend to partition into the epicuticular waxes covering many of the aboveground plant parts. While this may decrease leaching of the molecule from the plant surface by rain, it also influences ethylene release. The effect of lipid solubility on ethylene release is illustrated by the differential flux of ethylene evolved from (2-chloroethyl)phosphonic acid and (2-chloroethyl) methylbis(phenylmethoxy)silane following their application to the adaxial (upper) side of hemistomatous peach leaves (leaves with stomata only on the abaxial (lower) side) (Figure 8.)[144] Very little of the ethylene evolved from (2-chloroethyl)phosphonic acid (a hydrophilic molecule) migrated through the leaf to exit from the abaxial side. For (2-chloro-ethyl)methylbis(phenylmethoxy)silane, (a lipophilic molecule) abaxial flux of ethylene actually exceeded adaxial flux. The implication is that the latter compound partitions into the waxy portions of the leaf cuticle wherein breakdown and subsequent ethylene release take place. (2-Chloroethyl)phosphonic acid evidently remains chiefly on the surface, releasing ethylene into the air rather than into the interior of the leaf. Movement of hydrophobic (2-chloroethyl)methylbis(phenylmethoxy)silane parent molecules into cuticular and epicuticular waxes also has the potential to decrease the rate of breakdown and ethylene release since the parent molecule is stabilized by a nonpolar environment: This is illustrated by the difference in the rate of ethylene release from (2-chloroethyl)methylbis(phenylmeth-oxy)silane when applied to Teflon®, wax paper, and aluminum foil. The hydrophobic nature of the parent molecule permits penetration into the wax which decreases the rate of breakdown and ethylene release. This implies that concentration and composition of epicuticular waxes

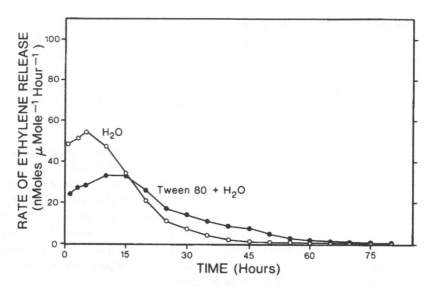

FIGURE 9. Effect of a surfactant (0.1% Tween® 80) on rate of ethylene release from (2-chloroethyl)methylbis(phenylmethoxy)silane on wax paper at 25°C.

on leaves, fruits, and flowers may alter the release of ethylene and the subsequent response of the plant.

2. Effect of Additives

Other chemicals applied concurrently or as an additional pre- or postethylene-releasing compound spray, may be used to modify the response desired. These include additives which alter the rate of breakdown of the ethylene-releasing compound, those which alter its surface coverage or penetration, and additives which modify the response of the crop to the ethylene released.

Surfactants such as X-77 (alkyl aryl polyethoxy ethanol and free fatty acids) and Tween® 20 (polyoxyethylene 20-sorbitan monolaurate) decrease the surface tension and increase the uniformity of coverage. This can improve the penetration of the parent molecule and as a consequence enhance both the uniformity and degree of the response by the plant.[56,59,108,109] Furthermore, addition of a surfactant such as Tween® 80 [polyoxyethylene (20) sorbitan monooleate] has the ability to slow the rate of ethylene release from (2-chloroethyl)methylbis(phenylmethoxy)silane (Figures 7 and 9). The surfactant possibly exerts its stabilizing effect by incorporating the ethylene-releasing compound into the hydrophobic portion of surfactant micelles. Glycerine, like surfactants, has been used to prolong the drying period and increase the penetration of (2-chloroethyl)phosphonic acid sprays.[145] The over threefold increase in penetration was accompanied by an increase in plant response relative to sprays without added glycerine. Other compounds which enhance plant response include the fungicide chlorothalonil (2,4,5,6-tetrachloroisothalonitrate) which appears to increase foliar uptake[110] and antitranspirants which are thought to decrease the rate of emanation of ethylene from the tissue.[69]

Chemical additives such as potassium iodide, bromodine, and urea have been used with (2-chloroethyl)phosphonic acid to increase abscission responses in some crops.[27,38,39] It is suggested that the effect is through increased uptake of the ethylene-releasing compound or action of the additive itself since no effect is apparent on the rate of ethylene release from the parent compound.[103] With some fruits (e.g., cherries[55]) additives may potentially decrease fruit quality, minimizing their usefulness.

Altering the release kinetics of ethylene by stabilizing or destabilizing the aqueous environment of the parent molecule at application is a practice used with some success. Alteration of the spray solution pH for (2-chloroethyl)phosphonic acid with citrate-dibasic sodium phosphate or potassium-phosphate is common.[111,112] As the pH progressively increases, the rate of breakdown increases which often enhances effectiveness.[67,112] Substances such as palm oil, however, used as a carrier for (2-chloroethyl)phosphonic acid applied to stimulate latex flow in rubber trees, decrease the rate and prolong the duration of ethylene release.[16,34,92]

With most crops, the target organ represents only a small portion of the aboveground plant. Seldom is it possible in a commercial setting to apply an ethylene-releasing compound to this site only. As a consequence, undesirable ethylene-mediated effects may occur in peripheral, nontarget organs. For example, leaf abscission during pecan and walnut dehiscence and olive fruit loosening is a common problem.[62,71,72,113] Application of calcium acetate sprays 2 weeks prior to the use of (2-chloroethyl)phosphonic acid on olives greatly decreases the degree of leaf loss.[62,70] Auxin or auxin-like substances are known to inhibit abscission.[114] In some situations, auxins may also counteract ethylene-mediated abscission resonses.[27,113] In apple ripening[58,115] and olive loosening,[27] combined or supplemental sprays of these compounds generally decrease leaf and fruit abscission.

B. Plant Factors

Plants vary widely in their susceptibility and response to exogenous ethylene. In addition to species variation, dose-response curves vary among different organs on the same plant (e.g., leaves, fruit, and roots), different cells within a specific organ, and in a few cases among different physiological responses occurring within the same cells.[116] Part of the variation in response to exogenous ethylene stems from the fact that the physiological response derived generally has little to do with the molecular mode of action of the hormone. These responses are many steps removed from the initial action of ethylene and are largely dependent upon the state of the cytoplasm at the time of exposure. Thus, factors that alter the physiological condition of the plant may also significantly alter the response obtained.

1. Species and Cultivar

Significant differences exist between species and in many cases between cultivars within a species in the response to ethylene-releasing compounds. Differences are found not only in the optimum concentration for a given response, but also in the acceptable range. For example, blossom thinning requires 300 to 500 ppm of (2-chloroethyl)phosphonic acid for peaches[43] while 1000 to 2000 ppm are needed for apples.[38] Cultivar differences may be equally pronounced. Application of 2250 to 2750 ppm of (2-chloroethyl)methyl-bis(phenylmethoxy)silane adequately thinned peach blossoms of the cultivar 'Candor', while sprays of only 1500 ppm overthinned the cultivar 'Jefferson'.[47]

Variation among species and cultivars has not been sufficiently studied at a mechanistic level. Response differences may reflect variation in the basic chemical and biochemical nature of the respective organs. It may also be in part due to morphological differences. Physical characteristics of the plant, such as cuticle thickness[108] and continuity,[105] leaf orientation, size and density, and stomate[105,117] and trichome[105] densities, are often important. Stomatal distribution can affect ethylene flux from an ethylene-releasing compound, but the effect is partly dependent upon the lipophilic nature of the parent molecule.[144] As noted previously, ethylene derived from (2-chloroethyl)phosphonic acid did not penetrate hemi-stomatous leaf tissue when the parent molecule was in contact with the stomate-free adaxial surface. Significant ethylene movement does occur, however, for amphistomatous leaf material when stomata are open (Figure 10). Ethylene from (2-chloroethyl) methylbis(phenylmethoxy)silane escapes from both surfaces in similar quantities regardless of the stomatal distribution (compare Figures 8B and 11A). Thus it appears that differences

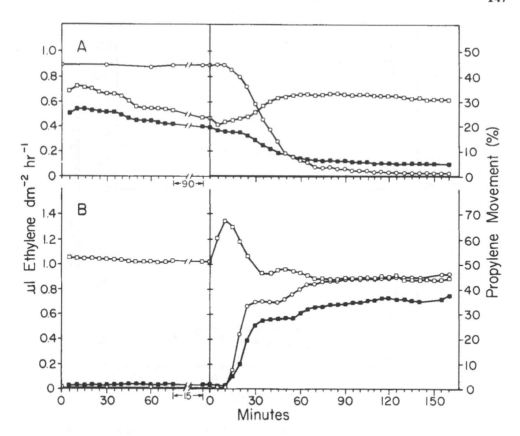

FIGURE 10. Time course of abaxial (■) and adaxial (□) ethylene flux and percent total propylene movement from the adaxial to abaxial side of the leaf (o) for amphistomatous pepper leaves treated adaxially with a 6.75 mM solution (pH 7) of (2-chloroethyl)phosphonic acid applied in light, then transferred to dark (A) and applied in dark, then transferred to light (B). Propylene was added to the adaxial leaf side and its appearance on the abaxial leaf side was used as a marker for stomatal opening.

in stomatal distribution will more likely affect plant responses to hydrophilic, rather than lipophilic, ethylene-releasing compounds.

Differences in the ability of the plant to metabolize a given ethylene-releasing compound may also contribute to variation in response among plants. Metabolic or chemical alterations in an ethylene-releasing compound may or may not alter its effectiveness. For example, the chloroethyl carbons of (2-chloroethyl)phosphonic acid may be to some extent converted to more stable compounds of unknown biological activity.[118] Metabolism of (2-chloroethyl)phosphonic acid has been found in squash,[119] cherry leaves[120,121] and rubber trees,[122-124] but not in peach,[125] cucumber,[118] tomato,[118] walnut,[126] olive,[127] citrus,[128] apple,[120] tobacco,[129,130] or cherry fruit.[120] Noncovalent complexes of (2-chloroethyl)phosphonic acid with sugars have been reported in peaches,[131] however, more recent extraction techniques pose a question as to the existence of these complexes in vivo.[122]

2. Physiological Status of the Plant

The physiological status of the plant or plant part can have a pronounced effect on its response to an ethylene-releasing compound. Plants tend to be much more sensitive to ethylene when under some form of stress. Thus, weakened trees or branches of sweet[53] and sour[55] cherries and French prunes[52] exhibit excessive leaf drop. Mineral deficiencies may likewise predispose the plant to undesirable secondary effects from the ethylene-releasing compound (e.g., leaf abscission).

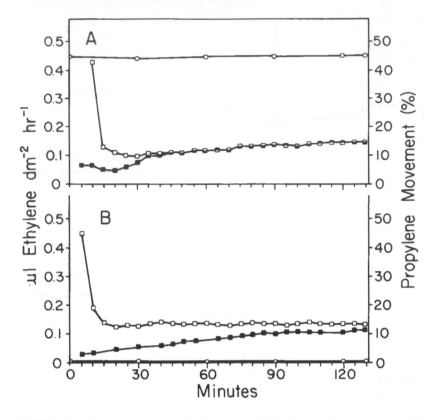

FIGURE 11.   Time course of abaxial (■) and adaxial (□) ethylene flux and percent total propylene movement from the adaxial to abaxial side of the leaf (o) for amphistomatous pepper leaves treated adaxially with a 6.75 mM solution (pH 7) of (2-chloroethyl) methylbis(phenylmethoxy)silane during light-induced stomatal opening (A) and dark-induced stomatal closure (B). Propylene was added to the adaxial leaf side and its appearance on the abaxial leaf side was used as a marker for stomatal opening.

Macro- and microenvironment strongly influences the conditions of the plant and thus its response to ethylene. Temperature not only modulates the rate of release of ethylene from the ethylene-releasing compound (Figure 3), but also significantly influences the responsiveness of the plant to ethylene. Low temperatures during or following application have been shown to delay and/or reduce the response by the plant.[27,57,63,73,132] Relative humidity has also been implicated in response variation.[132] For leaf material, some of the effects of environment may be related to the effects on stomatal aperture. Conditions of high vapor pressure deficit, low light, water stress, etc. may limit ethylene movement into the leaf following stomatal closure.[144] This effect is pronounced for (2-chloroethyl)phosphonic acid, which tends to remain on the surface of the leaf, but not for (2-chloroethyl) methylbis(phenylmethoxy)silane (compare Figures 10 and 11).

3. Maturity

The maturity of the plant or plant part often has a considerable influence on the effectiveness of an ethylene-releasing compound. For example, the sensitivity of young peach fruits to abscission changes markedly with age.[42,49,50] The concentration of (2-chloroethyl)methylbis(phenylmethoxy)silane required for optimum thinning decreased from 1500 to 2000 ppm at bloom[47,48] to 480 ppm when the fruit ovule length was 8.8 mm and 240 ppm when 13.9 mm.[49] Similarly, young peach leaves were less susceptible than older leaves to (2-chloroethyl)phosphonic acid and (2-haloethyl)silanes.[132] Unlike those of peach, younger cotton leaves appear to be more sensitive than mature, fully expanded leaves.[36] In both cotton and peach, leaves abscised while still green and had no outward appearance of undergoing senescence.

The effect of maturity on the magnitude of response to an ethylene-releasing compound is also seen in ripening. Climacteric fruits such as apple, banana, tomato, and avocado will not ripen until reaching a certain minimum stage of maturity.[35,83] Thus, application of an ethylene-releasing compound is only beneficial after reaching this point and the concentration required generally declines with advancing maturity.

V. PENETRATION AND TRANSLOCATION OF ETHYLENE-RELEASING COMPOUNDS

The penetration of ethylene-releasing compounds varies widely with the parent compound utilized, species and plant part treated, additives, and environmental conditions. In general, greater uptake has been found for (2-chloroethyl)phosphonic acid than (2-chloro-ethyl)methylbis(phenylmethoxy)silane, which in part probably reflects differences in water solubility.

Penetration of ^{14}C-chloroethyl labeled (2-chloroethyl)phosphonic acid varies widely among species[127,128] and in some cases cultivar.[109] When applied to cherry leaves, the abaxial surface exhibited 20 to 25 times more penetration than the adaxial surface.[109] The presence of light and the use of a surfactant (X-77) also significantly increased penetration but to a much lesser extent than temperature. Leaf penetration of (2-chloroethyl)phosphonic acid increased 55 times between 15 and 35°C, suggesting the possiblity of a temperature-dependent cuticular phase change being a dominant factor.

In contrast to cherry, little uptake was found in citrus leaves.[128] The compound tended to remain at the site of application and was translocated no more than 5 to 10 cm from the fruit and leaves into the subtending stem. In walnuts,[126] leaf penetration and subsequent translocation was significant in the leaves of very young seedling trees (2 to 3 true leaves) but much less in leaves of mature trees. In contrast to (2-chloroethyl)phosphonic acid, there appears to be very little penetration and transport of (2-chloroethyl)methyl-bis(phenylmethoxy)silane in the vegetative tissue of the peach.[133]

Penetration of ethylene-releasing compounds into fruit tissue tends to be lower than vegetative tissue. Generally, penetration of (2-chloroethyl)phosphonic acid ranged from very slight to approximately 20%.[120,126,134] An exception to this, however, would be olive fruits where 63% of the label applied was reportedly absorbed within 4 hr.[118]

Movement of (2-chloroethyl)methylbis(phenylmethoxy)silane into peach fruits appears to be very slight.[133] Over 99% of the applied label did not penetrate beyond the surface millimeter of tissue. This lack of penetration reflects, in part, the extremely low water solubility of the parent molecule.

Translocation of foliar-applied (2-chloroethyl)phosphonic acid appears to be largely in a source to sink direction via the phloem. As with penetration, the extent of transport varies widely among species. Application of labeled (2-chloroethyl)phosphonic acid to mature leaves of tomato, squash, cucumber, apple, sweet cherry, sour cherry, and walnut resulted in recovery of some label in developing leaves and/or fruits.[119,120,126] In tomato, a significant portion (~10%) of the applied label had been translocated into developing fruits by the 7th day after application.[119] Within 24 hr, 3 to 9% of the ^{14}C applied to squash had been translocated from the leaves to other parts of the plant.[119] Application of labeled (2-chloroethyl)phosphonic acid to leaves adjacent to fruits in apple and cherry results in increased activity in fruits.[120] Activity tended to increase in the fruit for 24 to 72 hr, declining subsequently as the compound was degraded.[120] By 25 days after application, only 0.3% of applied label was recovered in cucumber fruits.[119]

The ability to translocate (2-chloroethyl)phosphonic acid that penetrates foliar tissue appears to vary with leaf age and location on the plant. Very little transport was found from mature leaves of walnut[126] while leaves on young seedlings apparently transport the molecule readily throughout the plant. Application to mature leaves adjacent to a fruit gave a very limited activity within the hull, shell, and kernel. Within the kernel, activity increased until the 5th day following application and declined thereafter. In grapes, leaf application did not result in label accumulation within the fruit although label was found in the rachis.[134]

Translocation of foliar-applied (2-chloroethyl)methylbis(phenylmethoxy)silane in the peach appears to be very slight. Of the small amount found within vegetative tissue, there was an equal distribution between acropetal and basipetal transport.[133]

VI. DECOMPOSITION AND/OR METABOLISM OF ETHYLENE-RELEASING COMPOUNDS

The breakdown of the parent molecule of (2-chloroethyl)phosphonic acid and (2-chloroethyl)methylbis(phenylmethoxy)silane has been reported in both in vivo and in vitro systems.[106,122-124,135] The majority of the parent molecule of (2-chloroethyl)phosphonic acid breaks down forming ethylene, chlorine, and phosphoric acid. Decomposition appears to be simply a base-catalyzed reaction not involving enzymatic activity.[101,136,137] In vitro dechlorination of (2-chloroethyl)phosphonic acid forming (2-hydroxyethyl)phosphonic acid over a range of pH values has been reported.[138] The percent conversion is low (3 to 4%) but could account for possible secondary products found in some experiments.

In a model soil system, (2-chloroethyl)methylbis(phenylmethoxy)silane was found to decompose rapidly yielding ethylene, benzyl alcohol, and low levels of (2-chloroethyl)methyl(phenylmethoxy)chlorosilane and (dichloroethyl)(chloroethylmethylphenylmethoxy)disiloxane.[106] The latter two compounds apparently occur in side reactions during the breakdown of the parent molecule. Their presence would suggest that under some circumstances the initial decomposition step involves attack at one of the two methoxy-silicon bonds rather than at the chloroethyl position.

In peach orchard soils, the methoxy groups on the parent molecule of (2-chloroethyl)methylbis(phenylmethoxy)silane are largely metabolized by soil organisms to CO_2, while the ethyl group is primarily liberated as ethylene.[106] The methyl group on the parent molecule remains bound to the silicon atom, forming insoluble organosilicates; little is metabolized by soil microorganisms to CO_2. The rate of soil breakdown of the parent molecule is strongly pH dependent and is favored by soil pHs below 6.0.

In vivo breakdown and metabolism of (2-chloroethyl)methylbis(phenylmethoxy)silane in peach fruits differs somewhat from that occuring in a silicon dioxide environment.[135] In addition to ethylene, primary products were benzyl alcohol, benzylglucoside, and benzylmethylgluoside. When labeled in the ethyl position, 98.9% of the original label could be accounted for after 35 days. The majority (89%) was given off as ethylene with 9.04% remaining as extractable residue. Only a very small percentage was given off as CO_2 (0.22%) or found as structural carbohydrates (0.32%), indicating very limited metabolism of the chloroethyl position.

VII. FUTURE PROSPECTS FOR ETHYLENE-RELEASING COMPOUNDS IN AGRICULTURE

The future for the use of ethylene-releasing compounds in agriculture appears promising. New compounds with differing chemical characteristics and release kinetics should allow increased use of ethylene-releasing compounds for species which have more sensitive physiological responses and therefore require much more precise concentration control. In order to facilitate this expanded use in more critical situations, it is essential that we develop a greater understanding of the operative forces controlling both the rate of ethylene release from the parent molecule and the degree of response of various plant tissues and organs to ethylene.

REFERENCES

1. **Rodriguez, A.,** Influence of smoke and ethylene on the fruiting of the pineapple (*Ananas sativus* Shult.), *J. Dep. Agric. P.R.*, 26, 5, 1932.
2. **Gonzalez, L. G.,** The smudging of mango trees and its effects, *Philipp. Agric.*, 12, 15, 1924.
3. **Eaks, I. L.,** Ripening and astringency removal in persimmon fruit, *Proc. Am. Soc. Hortic. Sci.*, 91, 868, 1967.
4. **Sievers, A. F. and True, R. H.,** A preliminary study of the forced curing of lemons as practiced in California, *U.S. Dept. Agric., Bur. Plant Ind. Bull.*, 1, 1912.
5. **Cousins, H. H.,** *Annu. Rep. Dep. Agric. Jamaica*, Vol. 7, 1910.
6. **Neljubov, D.,** Uber die horizontale Nutation der Stengel von *Pisum sativum* und einiger anderen Pflanzen, *Beih. Bot. Centralkl.*, 4, 60, 1901.
7. **Knight, L. L. and Crocker, W.,** Toxicity of smoke, *Bot. Gaz.*, 55, 337, 1913.
8. **Gane, R.,** Production of ethylene by some ripening fruits, *Nature (London)*, 134, 1008, 1934.
9. **Hills, L. D. and Haywood, E. H.,** *Rapid Tomato Ripening*, Faber & Faber, London, 1946, 143.
10. **Gowing, D. P. and Leeper, R. W.,** Induction of flowering in pineapple by beta-hydroxyethylhydrazine,
11. **Palmer, R. L., Lewis, L. N., Hield, H. Z., and Kumamoto, J.,** Abscission induced by betahydroxy-ethylhydrazine: conversion of betahydroxyethylhydrazine to ethylene, *Nature*, 216, 1216, 1967.
12. **Kumamoto, J., Dollwet, H. H. A., and Lyons, J. M.,** Evidence and hypothesis for a "Taube Bridge Electron Transfer" propagating to a remote site through δ bonding. The formation of ethylene from monoethyl sulfate, *J. Am. Chem. Soc.*, 91, 1207, 1969.
13. **Dollwet, H. H. A. and Kumamoto, J.,** Ethylene production of ethyl propylphosphate, Niagra 10637, *Plant Physiol.*, 46, 786, 1970.
14. **Kabachnik, M. I. and Rossiiskaya, P. A.,** Organophosphorous compounds. I. Reaction of ethylene oxide with phosphorus trichloride, *Izv. Akad. Nauk. SSSR, Otdel. Khim. Nauk*, 406, 295, 1946.
15. **Maynard, J. A. and Swan, J. M.,** Organophosphous compounds. I. 2-chloroalkylphosphonic acid as phosphorylating agents, *Aust. J. Chem.*, 16, 596, 1963.
16. **Hofer, W. and Luersson, K.,** Novel Halogenoethyl Sulphones and Their Use as Plant Growth Regulators, U.S. Patent 4,227,918, 1978.
17. **Lürssen, K.,** Manipulation of crop growth by ethylene and some implications of the mode of generation, *Easter Sch. Agric. Sci. Univ. Notthingham, Proc.* 33, 66, 1982.
18. **Yu, Y. B., Adams, D. O., and Yang, S. F.,** 1-Aminocyclopropanecarboxylate synthase, a key enzyme in ethylene biosynthesis, *Arch. Bioch. Biophys.*, 198, 280, 1979.
19. **Guy, M. and Kende, H.,** Ethylene formation in *Pisum sativum* and *Vicia faba* protoplasts, *Planta*, 160, 276, 1984.
20. **Bradford, K. J. and Yang, S. F.,** Xylem transport of 1-aminocyclopropane-1-carboxylic acid, an ethylene precursor, in waterlogged tomato plants, *Plant Physiol.*, 65, 322, 1980.
21. **Hume, B. and Lovell, P.,** Role of aminocyclopropane-1-carboxylic acid in ethylene release by distal tissues following localized application of ethephon in *Cucurbita pepo*, *Physiol. Plant.*, 58, 101, 1983.
22. **Lavee, S. and Martin, G. C.,** Ethylene evolution following treatment with 1-aminocyclopropane-1-carboxylic acid and ethephon in an *in vitro* olive shoot system in relation to leaf abscission, *Plant Physiol.*, 67, 1204, 1981.
23. **Zimmerman, P. W. and Wilcoxon, F.,** Several chemical growth substances which cause initiation of roots and other responses in plants, *Contrib. Boyce Thompson Inst.*, 7, 209, 1935.
24. **Yoshii, H. and Imaseki, H.,** Biosynthesis of auxin-induced ethylene. Effects of indole-3-acetic acid, benzyladenine and abscisic acid on endogenous levels of 1-aminocyclopropane-1-carboxylic acid (ACC) and ACC synthase, *Plant Cell Physiol.*, 22, 369, 1981.
25. **Morgan, P. W. and Hall, W. C.,** Accelerated release of ethylene by cotton following application of indolyl-3-acetic acid, *Nature*, 201, 99, 1964.
26. **Morgan, P. W.,** Effects on ethylene physiology, in *Herbicides: Physiology, Biochemistry, Ecology*, Vol. 1, 2nd ed., Audus, L. J., Ed, Academic Press, New York, 1976, 255.

27. **Hartmann, H. T., Tombesi, A., and Whisler, J.,** Promotion of ethylene evolution and fruit abscission in the olive by 2-chloroethylphosphonic acid and cycloheximide, *J. Am. Soc. Hortic. Sci.,* 95, 635, 1970.

28. **Saltveit, M. E., Jr., and Dilly, D. R.,** Rapidly induced wound ethylene from excised segments of etiolated *Pisum sativum* L. cv. Alaska. I. Characterization of the response, *Plant Physiol.,* 61, 447, 1978.

29. **Goeschl, J. D., Rappaport, L., and Pratt, H. A.,** Ethylene as a factor regulating the growth of pea epicotyls subjected to physical stress, *Plant Physiol.,* 41, 877, 1966.

30. **Hoffman, N. E. and Yang, S. F.,** Enhancement of wound-induced ethylene synthesis by ethylene in preclimacteric cantaloupe, *Plant Physiol.,* 69, 317, 1982.

31. **Copper, W. C. and Henry, W. H.,** Field trials with potential abscission chemicals as an aid to mechanical harvesting of citrus in Florida, *Proc. Fla. State Hortic. Soc.,* 81, 62, 1968.

32. **Cooper, C. C. and Henry, W. H.,** Abscission chemicals in relation to citrus fruit harvest, *J. Agric. Food Chem.,* 19, 559, 1971.

33. **Archer, B. A.,** Yield stimulants for *Hevea brasiliensis, Chem. Ind.,* 7, 282, 1974.

34. **Bridge, K.,** Plant growth regulator use in natural rubber (*Hevea brasiliensis*), in *Plant Growth Regulating Chemicals,* Volume I, Nickell, L. G., Ed., CRC Press, Boca Raton, Fla., 1983, 41.

35. **Abeles, F. A.,** *Ethylene in Plant Biology,* Academic Press, N.Y., 1973, 302.

36. **Morgan, P. W.,** Stimulation of ethylene evolution and abscission in cotton by 2-chloroethanephosphonic acid, *Plant Physiol.,* 44, 337, 1969.

37. **de Wilde, R. C.,** Practical applications of (2-chloroethyl)phosphonic acid in agricultural production, *HortScience,* 6, 364, 1971.

38. **Amchem Products, Inc.,** Ethrel, *Technical Service Data Sheet H-96,* Ambler, Pa., 1969, 64.

39. **Larsen, F. E.,** Promotion of leaf abscission of deciduous nursery stock with 2-chloroethylphosphonic acid, *J. Am. Hortic. Sci.,* 95, 662, 1970.

40. **Larsen, F. E.,** Potassium iodide induced leaf abscission of deciduous woody plants, *Proc. Am. Soc. Hortic. Sci.,* 88, 690, 1966.

41. **Larsen, F. E.,** Pre-storage promotion of leaf abscission of deciduous nursery stock with bromodine, *J. Am. Soc. Hortic. Sci.,* 95, 231, 1970.

42. **Stembridge, G. E. and Gambrell, G. E., Jr.,** Thinning peaches with bloom and postbloom applications of 2-chloroethylphosphonic acid, *J. Am. Soc. Hortic. Sci.,* 96, 7, 1971.

43. **Buchanan, D. W., and Biggs, R. H.,** Peach fruit abscission and pollen germination as influenced by ethylene and 2-chloroethane-phosphonic acid, *J. Am. Soc. Hortic. Sci.,* 94, 327, 1969.

44. **Daniell, J. W. and Wilkinson, R. E.,** Effect of ethephon-induced ethylene on abscission of leaves and fruits of peaches, *J. Am. Soc. Hortic. Sci.,* 97, 682, 1972.

45. **Buchanan, D. W., Biggs, R. H., Blake, J. A., and Sherman, W. B.,** Peach thinning with 3CPA and ethrel during cytokinesis, *J. Am. Soc. Hortic. Sci.,* 95, 781, 1970.

46. **Blake, J. A., Biggs, R. H., and Buchanan, D. W.,** Post-bloom thinning of Florida peaches with 2-chloroethylphosphonic acid, *Fla. State Hortic. Soc.,* 82, 257, 1969.

47. **Krewer, G. W., Daniell, J. W., and Couvillon, G. A.,** Peach blossom thinning with CGA-15281, *HortScience,* 17, 663, 1982.

48. **Arnold, C. E., Aldrich, J. H., and Smith, H. R.,** Blossom thinning of peach with CGA-15281, *Proc. Fla. State Hortic. Soc.,* 95, 128, 1982.

49. **Dozier, W. A., Jr., Carlton, C. C., and Short, K. C.,** Thinning 'Loring' peaches with CGA-15281, *HortScience,* 16, 56, 1981.

50. **Gambrell, C. E., Jr., Coston, D. C., and Sims, E. T., Jr.,** Results of eight years with CGA-15281 as a postbloom thinner for peaches, *J. Am. Soc. Hortic. Sci.,* 108, 605, 1983.

51. **Martin, G. C., Fitch, L. B., Carnill, G. L., and Ramos, D. E.,** Thinning French prune (*Prunus domestica* L.) with (2-chloroethyl)phosphonic acid, *J. Am. Soc. Hortic.,* 100, 90, 1975.

52. **Sibbett, G. S. and Martin, G. C.,** Cumulative effects of ethephon as a fruit thinner on French prune (*Prunus domestica* L.), *HortScience,* 17, 665, 1982.

53. **Bukovac, M. J.,** Machine-harvest of sweet cherries: effect of ethephon on fruit removal and quality of the processed fruit, *J. Am. Soc. Hortic. Sci.,* 104, 289, 1979.

54. **Bukovac, M. J., Zucconi, F., Wittenbach, V. A., Flore, J. A., and Inoue, H.,** Effects of (2-chloroethyl)phosphonic acid on development and abscission of maturing sweet cherry (*Prunus avium* L.) fruit, *J. Am. Soc. Hortic. Sci.,* 96, 777, 1971.

55. **Bukovac, M. J., Zucconi, F., Larsen, R. P., and Kessler, C. D.,** Chemical promotion of fruit abscission in cherries and plums with special reference to 2-chloroethylphosphonic acid, *J. Am. Soc. Hortic. Sci.,* 94, 226, 1969.

56. **Wittenbach, V. A. and Bukovac, M. J.,** An anatomical and histochemical study of abscission in maturing sweet cherry fruit, *J. Am. Soc. Hortic. Sci.,* 97, 214, 1972.

57. **Olien, W. C. and Bukovac, M. J.,** Ethylene generation, temperature responses, and relative biological activities of several compounds with potential for promoting abscission of sour cherry fruit, *J. Am. Soc. Hortic. Sci.,* 107, 1085, 1982.

58. **Edgerton, L. J. and Blanpied, G. D.**, Interaction of succinic acid 2,2-dimethyl hydrazide, 2-chloro-ethylphosphonic acid and auxins on maturity, quality and abscission of apples, *J. Am. Soc. Hortic. Sci.*, 95, 664, 1970.

59. **Rasmussen, G. K.**, The effect of additives on the efficiency of abscission-inducing chemicals on 'Valencia' oranges in Florida, *Proc. Fla. State Hortic. Soc.*, 93, 27, 1980.

60. **Rasmussen, G. K.**, Loosening of oranges with Pik-Off, Release, Acti-Aid, and Sweep combinations, *Proc. Fla. State Hortic. Soc.*, 90, 4, 1977.

61. **Rasmussen, G. K.**, Results of 5 years continued use of abscission-inducing chemicals on 'Hamlin' oranges, *Proc. Fla. State Hortic. Soc.*, 92, 51, 1979.

62. **Martin, G. C., Lavee, S., and Sibbett, G. S.**, Chemical loosening agents to assist mechanical harvest of olive, *J. Am. Soc. Hortic. Sci.*, 106, 325, 1981.

63. **Sun, F. Z. and Martin, G. C.**, Evaluation of (2-chloroethyl)methylbis(phenylmethoxy)silane (CGA-15281) as a chemical fruit abscising agent for olive using detached shoots, *HortScience*, 17, 957, 1982.

64. **Vitagliano, C.**, Ethylene-releasing compounds to loosen olive fruit for mechanical harvesting, *HortScience*, 10, 591, 1975.

65. **Vitagliano, C.**, Effects of ethephon on stomata, ethylene evolution, and abscission in olive (*Olea europaea* L.), cv. Coratina, *J. Am. Soc. Hortic. Sci.*, 100, 482, 1975.

66. **Rufener, J. and Green, D. H.**, Alsol, a new chemical harvest aid for olives, *Plant Growth Regulator, Proc. 2nd Int. Symp.*, 372, 1975.

67. **Forlani, M., Pugliano, G., and Rotundo, A.**, Mechanical harvesting of olives: use of three "ethylene-promoters" on the Carolea cultivar, *Ann. Fac. Sci. Agrar. Univ. Studi Napoli, Portici*, 9-10, 237, 1976 (abstract).

68. **Rotundo, A., and Pasquarella, C.**, The use of ethylene products in the mechanical harvesting of olives. Part 2, *Ann. Fac. Sci. Agrar. Univ. Studi. Napoli, Portici.*, 12(Abstr.), 6, 1978.

69. **El-Tamzini, M. I., Shaladan, M. S., and Niazi, Z. M.**, Antitranspirant increases Alsol effectiveness as chemical aid for harvesting olive, *HortScience*, 17, 965, 1982.

70. **El-Tamzini, M. I., Niazi, Z. M., and Shaladan, M. S.**, Calcium reduces defoliation induced by 2-chloroethyl-tris-(2-methoxyethoxy)silane in olive, *HortScience*, 17, 966, 1982.

71. **Wood, B. W.**, Fruit thinning of pecan with ethephon, *HortScience*, 18, 53, 1975.

72. **Kays, S. J., Crocker, T. F., and Worley, R. E.**, Concentration dependencies of ethylene on shuck dehiscence and fruit and leaf abscission of *Carya illinoensis* (Wang.) K. Koch, *J. Agric. Food Chem.*, 23, 1116, 1975.

73. **Pratt, H. K. and Goeschl, J. D.**, Physiological roles of ethylene in plants, *Annu. Rev. Plant Physiol.*, 20, 541, 1969.

74. **Dozier, W. A., Jr. and Bardens, J. A.**, Shoot growth of young apple trees as influenced by (2-chloroethyl)phosphonic acid, *J. Am. Soc. Hortic. Sci.*, 98, 244, 1973.

75. **Szyjewicz, E. and Kliewer, W. M.**, Influence of temperature and ethephon concentration on growth and composition of Cabernet Sauvignon grapevines, *J. Plant Growth Regul.*, 1, 295, 1982.

76. **Cooke, A. R. and Randall, D. I.**, 2-haloethanephosphonic acids as ethylene releasing agents for the induction of flowering in pineapples, *Nature*, 218, 974, 1968.

77. **Edgerton, L. J. and Greenhalgh, W. J.**, Regulation of growth, flowering and fruit abscission of apples and peaches with Amchem 66-329, *J. Am. Soc. Hortic. Sci.*, 94, 11, 1969.

78. **Williams, M. W.**, Postharvest application of growth regulators to increase spur formation and subsequent flowering and fruiting of young apple trees, *HortScience*, 5, 349, 1970.

79. **Dennis, F. G., Jr.**, Trials of ethephon and other growth regulators for delaying bloom in tree fruits, *J. Am. Soc. Hortic. Sci.*, 101, 241, 1976.

80. **McMurray, A. L. and Miller, C. H.**, Cucumber sex expression modified by 2-chloroethanephosphonic acid, *Science*, 162, 1397, 1968.

81. **Miller, C. H. and McMurray, A. L.**, Some effects of Ethrel (2-chloroethanephosphonic acid) on vegetable crops, *HortScience*, 4, 248, 1969.

82. **Shannon, S. and Robinson, R. W.**, The use of ethephon to regulate sex expression of summer squash for hybrid seed production, *J. Am. Soc. Hortic. Sci.*, 104, 674, 1979.

83. **Burg, S. P. and Burg, E. A.**, Ethylene action and the ripening of fruit, *Science*, 148, 1190, 1965.

84. **Robinson, R. W., Wilczynski, H., Dennis, F. D. and Bryan, H. H.**, Chemical promotion of tomato fruit ripening, *Proc. Am. Soc. Hortic. Sci.*, 93, 823, 1968.

85. **Edgerton, L. J. and Blanpied, G. D.**, Regulation of growth and fruit maturation with 2-chloroethane-phosphonic acid, *Nature*, 219, 1064, 1968.

86. **Ben-Yehoshua, S., Iwamori, S., and Lyons, J. M.**, Ethylene and 2-chloroethanephosphonic acid: role in development of fig fruit, *Am. Soc. Hortic. Sci. Symp.*, 1968.

87. **Shawa, A. Y.**, Effect of ethephon on color, abscission, and keeping quality of 'Mcfarlin' cranberry, *HortScience*, 14, 168, 1979.

88. **Devlin, R. M. and Demoranville, I. E.**, Influence of two ethylene-releasing compounds of anthocynanin formation, size, and yield in 'Early Black' cranberries, *Proc. N.E. Weed Sci. Soc.*, 32, 108, 1978.

89. **Russo, L., Jr., Dostal, H. C., and Leopold, A. C.**, Chemical regulation of fruit ripening, *BioScience*, 18, 109, 1968.

90. **Nickell, L. G.**, Sugarcane, in *Plant Growth Regulating Chemicals*, Volume I, Nickell, L. G., Ed., CRC Press, Boca Raton, Fla., 1983, 185.

91. **Steffens, G. L.**, Tobacco, in *Plant Growth Regulating Chemicals*, Volume I, Nickell, L. G., Ed., CRC Press, Boca Raton, Fla., 1983, 71.

92. **Abraham, P. D., Wycherley, P. R., and Pakianathan, S. W.**, Stimulation of latex flow in *Hevea brasiliensis* by 4-amino-3,4,6-trichloropicolinic acid and 2-chloroethanephosphonic acid, *J. Rubber Res. Inst. Malays.*, 20, 291, 1968.

93. **Foery, W. and Fischer, H. P.**, β-Halogenoethyl silanes as stimulators for latex flow, U.S. Patent 4,004,908, 1977.

94. **Stahmann, M. A., Clare, G. B., and Woodbury, W.**, Increased disease resistance and enzyme activity induced by ethylene and ethylene production by black rot infected sweet potato tissue, *Plant Physiol.*, 41, 1505, 1966.

95. **Kamal, A. L. and Marroush, M.**, Control of chocolate spot in potato tubers by foliar spray with 2-chloroethylphosphonic acid, *HortScience*, 6, 42, 1971.

96. **Liptay, A., Phatak, S. C., and Jaworski, C. A.**, Ethephon treatment of tomato transplants improves frost tolerance, *HortScience*, 17, 400, 1982.

97. **Proebsting, E. L., Jr. and Mills, H. H.**, Effects of growth regulators on fruit bud hardiness in *Prunus*, *HortScience*, 4, 254, 1969.

98. **Worku, Z., Herner, R. C., and Carolus, R. L.**, Effect of stage of ripening and ethephon treatment on color content of paprika pepper, *Sci. Hortic.*, 3, 239, 1975.

99. **Furuta, T., Humphrey, W., Maire, R., and Yamamato, Y.**, Controlling fruit formation on olive and Victorian box with Off-Shoot-O and Ethrel, *Calif. Agric.*, 24, 11, 1970.

100. **Pakianathan, S. W.**, Some factors affecting yield response to stimulation with 2-chloroethylphosphonic acid, *J. Rubber Res. Inst. Malays.*, 25, 50, 1977.

101. **Yang, S. F.**, Ethylene evolution from 2-chloroethylphosphonic acid, *Plant Physiol.*, 44, 1203, 1969.

102. **Warner, H. L. and Leopold, A. C.**, Ethylene evolution from 2-chloroethylphosphonic acid, *Plant Physiol.*, 44, 156, 1969.

103. **Biddle, E., Kerfoot, D. G. S., Kho, Y. H., and Russell, K. E.**, Kinetic studies of the thermal decomposition of 2-chloroethylphosphonic acid in aqueous solution, *Plant Physiol.*, 58, 700, 1976.

104. **Klein, I., Lavee, S., and Ben-Tal, Y.**, Effect of water vapor pressure on the thermal decomposition of 2-chloroethylphosphonic acid, *Plant Physiol.*, 634, 474, 1979.

105. **Sargent, J. A.**, The penetration of growth regulators into leaves, in *Transport of Plant Hormones*, Vardar, Y., Ed., North-Holland, Amsterdam, 1968, 365.

106. **Kays, S. J., Arrendale, R. F., Seeley, S. D., and Couvillon, G. A.**, Decomposition and metabolism of the ethylene-releasing compound CGA-15281 in peach orchard soils, *J. Am. Soc. Hortic. Sci.*, 108, 661, 1983.

107. **Adams, D. O. and Yang, S. F.**, Ethylene biosynthesis: identification of 1-aminocyclopropane-1-carboxylic acid as an intermediate in the conversion of methionine to ethylene, *Proc. Natl. Acad. Sci. U.S.A.*, 76, 170, 1979.

108. **Westwood, M. N. and Batjer, L. P.**, Factors influencing absorption of dinitro-ortho cresol and napthaleneacetic acid by apple leaves, *Proc. Am. Soc. Hortic. Sci.*, 72, 35, 1958.

109. **Flore, J. A. and Bukovac, M. J.**, Factors influencing absorption of ^{14}C(2-chloroethyl)phosphonic acid by leaves of cherry, *J. Am. Soc. Hortic. Sci.*, 107, 965, 1982.

110. **Gilbert, M.**, Fate of chlorothalonil in apple foliage and fruit, *J. Agric. Food Chem.*, 24, 1004, 1978.

111. **Olien, W. C. and Bukovac, M. J.**, The effect of temperature on rate of ethylene evolution from ethephon and from ethephon-treated leaves of sour cherry, *J. Am. Soc. Hortic. Sci.*, 103, 199, 1978.

112. **Ben-Tal, Y. and Lavee, S.**, Increasing the effectiveness of ethephon for olive harvesting, *HortScience*, 11, 489, 1976.

113. **Martin, G. C.**, 2-Chloroethylphosphonic acid as an aid to mechanical harvesting of English wanluts, *J. Am. Soc. Hortic. Sci.*, 96, 434, 1971.

114. **Rubinstein, B. and Leopold, A. C.**, Auxin control of bean leaf abscission, *Plant Physiol.*, 38, 262, 1963.

115. **Crassweller, R. M.**, Effect of CGA-15281, an ethylene-generating material on maturity of 'Delicious' apple, *HortScience*, 17, 656, 1982.

116. **Goeschl, J. D. and Kays, S. J.**, Concentration dependencies of some effects of ethylene on etiolated pea, peanut, bean and cotton seedlings, *Plant Physiol.*, 55, 670, 1975.

117. **Barmore, C. R. and Biggs, R. H.**, Ethylene diffusion through citrus leaf and fruit tissue, *J. Am. Soc. Hortic. Sci.*, 97, 24, 1972.

118. **Giulivo, C., Ramina, A., Masia, A., and Costa, A.,** Metabolism and translocation of 1,2 (2-chloro-ethyl)phosphonic acid in *Prunus persica* (L.) Batsch, *Sci. Hortic.,* 15, 33, 1981.

119. **Yamaguchi, M., Chu, C. W., and Yang, S. F.,** The fate of ^{14}C(2-chloroethyl)phosphonic acid in summer squash, cucumber, and tomato, *J. Am. Soc. Hortic. Sci.,* 96, 606, 1971.

120. **Edgerton, L. J. and Hatch, A. H.,** Absorption and metabolism of ^{14}C(2-chloroethyl)phosphonic acid in apples and cherries, *J. Am. Soc. Hortic. Sci.,* 97, 112, 1972.

121. **Gilbert, D. M., Monselise, S. D., Edgerton, L. J., Maylin, G. A., Hicks, L. J., and Lisk, D. J.,** Metabolism studies with ethephon in cherry leaves, *J. Agric. Food Chem.,* 23, 290, 1975.

122. **Audley, B. G. and Wilson, H. M.,** Metabolism of 2-chloroethylphosphonic acid (ethephon) in suspension cultures of *Hevea brasiliensis, J. Exp. Bot.,* 29, 1329, 1978.

123. **Audley, B. G., Archer, B. L., and Carruthers, I. B.,** Metabolism of ethephon (2-chloroethylphosphonic acid) and related compounds in *Hevea brasiliensis, Arch. Environ. Contam. Toxicol.,* 4, 183, 1976.

124. **Archer, B. L., Audley, B. G., and Mann, N. P.,** Decomposition of 2-chloroethylphosphonic acid in stems and leaves of *Hevea brasiliensis, Phytochemistry,* 12, 1535, 1973.

125. **Lavee, S. and Martin, G. C.,** Ethephon [1,2-^{14}C(2-chloroethyl)phosphonic acid] in peach (*Prunus persica* L.) fruits. III. Stability and persistance, *J. Am. Soc. Hortic. Sci.,* 100, 28, 1975.

126. **Martin, G. C., Abdel Gawad, H. A., and Weaver, R. J.,** The movement and fate of (2-chloro-ethyl)phosphonic acid in walnut, *J. Am. Soc. Hortic. Sci.,* 97, 51, 1972.

127. **Epstein, E., Klein, I., and Lavee, S.,** The fate of 1,2^{14}C-(chloroethyl)phosphonic acid (ethephon) in olive (*Olea europea*), *Physiol. Plant.,* 39, 33, 1977.

128. **Young, R. H., and Jahn, O. L.,** The fate of 1,2-^{14}C-(2-chloroethyl)phosphonic acid in citrus, *J. Am. Soc. Hortic. Sci.,* 100, 496, 1975.

129. **Domir, S. C. and Foy, C. L.,** A study of ethylene and CO_2 evolution from ethephon in tobacco, *Pestic. Biochem. Physiol.,* 9, 1, 1978.

130. **Domir, S. C. and Foy, C. L.,** Movement and metabolic fate of [^{14}C]ethephon in flue-cured tobacco, *Pestic. Biochem. Physiol.,* 9, 9, 1978.

131. **Lavee, S. and Martin, G. C.,** Ethephon [1,2-^{14}C(2-chloroethyl)phosphonic acid] in peach fruits. II. Metabolism, *J. Am. Soc. Hortic. Sci.,* 99, 100, 1974.

132. **Porpiglia, P. J. and Barden, J. A.,** Peach leaf abscission following CGA-15281 and CGA-17856 applications as affected by temperature, *J. Am. Soc. Hortic. Sci.,* 105, 227, 1980.

133. **Couvillon, G. A., Seeley, S. D., and Kays, S. J.,** Uptake and translocation of [^{14}C-ethyl] labeled (2-chloroethyl)methylbis(phenylmethoxy)silane [CGA-15281] in the peach, *J. Am. Soc. Hortic. Sci.,* 107, 863, 1982.

134. **Weaver, R. J., Abdel-Gawad, H. A., and Martin, G. C.,** Translocation and persistence of 1,2-^{14}C-(2-chloroethyl)phosphonic acid (ethephon) in Thompson Seedless grapes, *Physiol. Plant.,* 26, 13, 1972.

135. **Seeley, S. D., Couvillon, G. A., and Kays, S. J.,** Metabolism of an ethylene-releasing growth regulator (CGA-15281) in young peach fruit, *J. Am. Soc. Hortic. Sci.,* 107, 682, 1982.

136. **Dennis, F. G., Wilczynski, H., de la Guardia, M., and Robinson, R. W.,** Ethylene leaves in tomato fruits following treatment with Ethrel, *HortScience,* 5, 168, 1970.

137. **Lougheed, E. C. and Franklin, E. W.,** Ethylene evolution from 2-chloroethylphosphonic acid under nitrogen atmospheres, *Can. J. Plant Sci.,* 50, 586, 1970.

138. **Audley, B. G. and Archer, B. L.,** Decomposition of 2-chloroethylphosphonic acid in aqueous solution: formation of 2-hydroxyethylphosphonic acid, *Chem. Ind.,* 13, 634, 1973.

139. **Kays, S. J., Seeley, S. D., and Couvillon, G. A.,** Uptake, transport and metabolism of the ethylene releasing compound CGA-15281 in the peach and metabolism in the soil, *Acta Hortic.,* 137, 29, 1983.

140. **Perry, S. C. and Gianfagna, T. J.,** Effect of Ethrel and Silaid on peach leaf and fruit abscission in relation to the kinetics of ethylene release, *Acta Hortic.,* 201, 157, 1987.

141. **Lang, G. A. and Martin, G. C.,** Ethylene-induced olive organ abscission: Ethylene pulse treatments improve fruit to leaf abscission ratios, *Acta Hortic.,* 201, 43, 1987.

142. **Lang, G. A. and Martin, G. C.,** Ethylene-releasing compounds and the laboratory modeling of olive fruit abscissions vs. ethylene release, *J. Am. Soc. Hortic. Sci.,* 110, 201, 1985.

143. **Beaudry, R. M. and Kays, S. J.,** Effects of physical and environmental factors on the release kinetics of ethylene from (2-chloroethyl)phosphonic acid and (2-chloroethyl)methylbis(phenylmethoxy)silane, *J. Am. Soc. Hortic. Sci.,* 112, 352, 1987.

144. **Beaudry, R. M. and Kays, S. J.** Factors affecting flux of ethylene from leaves treated with (2-chloro-ethyl)phosphonic acid and (2-chloroethyl)methylbis(phenylmethoxy)silane, *J. Am. Soc. Hortic. Sci.,* (in press).

145. **Ben-Tal, Y.,** Improving ethephon's effect on olive fruit abscission by glycerine, *HortScience,* 22, 869, 1987.

Chapter 9

GLOBAL AND HORTICULTURAL CARBON DIOXIDE ENRICHMENT

Herbert Zvi Enoch

TABLE OF CONTENTS

I. INTRODUCTION

Rates of penetration of CO_2 into the interior of plant leaves and its subsequent photosynthetic fixation into organic material are greatly influenced by CO_2 concentration gradients between the atmosphere and leaf cells. Mankind is presently performing — albeit unwillingly — the largest field experiment ever undertaken. Anthropogenic activities, largely burning of fossil fuel and deforestation, are increasing the global atmospheric CO_2 concentration by more than one third of a percent per year. In the last 100 years the global CO_2 concentration has increased from a preindustrial level believed to be about 280 $\mu\ell$ of CO_2 per liter day atmospheric air (μll^{-1}) to the present 340 μll^{-1}. Further increases are expected to lead to a doubling or even a severalfold increase of atmospheric CO_2 during the next one or two centuries. These elevated CO_2 concentrations are expected to affect plant productivity and modify the global climate. In this chapter an overview of climatological and plant processes which are likely to be affected is presented and the implications of the changes in atmospheric CO_2 concentration for photosynthesis and plant productivity are discussed.

Elevated levels of atmospheric CO_2 are used in greenhouses to increase the rate of production. The use of different types of CO_2 enrichment and the potential for CO_2 enrichment in different climatic zones are described. The use of methods to determine the CO_2 uptake efficiency, i.e., the fraction of CO_2 supplied to the greenhouse that is being assimilated by the crop, is discussed.

II. INTERACTIONS BETWEEN THE AERIAL ENVIRONMENT AND PLANT GROWTH

Plants need a suitable environment in order to grow well. If we subdivide the aerial environment into components that are of importance to plants, light, temperature, carbon dioxide, and oxygen concentration appear to be the main factors influencing productivity of terrestrial plants.

The amount of light during the daytime and the temperature in the tropical and subtropical zones and during the summertime in the temperate zone are relatively close to the optimum for plants. However, it is thought-provoking that for maximum growth rate, the CO_2 concentration is one order of magnitude too small and the O_2 concentration is one order of magnitude too big. In the following, it will be argued that plants are not the "innocent bystanders" which have to live in a hostile world with an uncomfortable CO_2 and O_2 composition, but can be looked upon as the major factor that has depleted the atmospheric CO_2 and caused an accumulation of O_2 that in turn, acts as an inhibitor of photosynthetic carbon fixation. Thus, plants have decreased potential plant productivity. Moreover, it will be argued that anthropogenic activities, such as burning of fossil fuel — some of which remains airborne — adds CO_2 to the global atmosphere and thus improves plant productivity.

III. THE HISTORY OF THE ATMOSPHERIC CARBON DIOXIDE OF THE EARTH

Green plants exchange CO_2 and O_2 with the global atmosphere in the photosynthesis process. The leaves of autotrophic terrestrial plants incorporate carbon from atmospheric CO_2 and fix energy from solar radiation. Part of that energy is used to split water — absorbed by the plant through its roots — into O_2 and H_2. The O_2 is emitted to the atmosphere and the H_2 is combined with carbon to form organic material. This organic material forms the basis for the food chain of heterotrophic organisms. The rates of photosynthesis, growth, development, and yield of plants are influenced by the interaction between atmospheric CO_2 concentration and their leaves. Consequently, all plants and animals are influenced by the CO_2 concentration in the bulk air.

Plants not only adapt passively to the atmospheric environment. To a large degree they create their environment by changing the global atmospheric composition and thereby the global surface temperature.

Earth has had a solid surface of rocks for about 4000 million years. Accordingly to Margulis,[1] the oldest fossils — unicellular photosynthesizing organisms resembling modern bacteria — existed about 3500 million years ago. Multicelled organisms only appeared some 600 million years ago, and consequently, larger fossils can only be found in rocks from after that period. During the three eons, unicellar organisms were the only life forms; in the words of Margulis,[1] "microbes had developed all the major biological adaptions: diverse energy-transforming and feeding strategies, movement, sensing, sex and even cooperation and competiton."

But not only did internal changes take place in organisms, these life forms to a large degree changed the physical environment of the Earth by changing the atmospheric composition and thereby the surface temperature. It is believed that photosynthesizing microbes decreased the global atmospheric CO_2 concentration and added O_2 to the atmosphere.[1]

According to Lovelock,[2] the atmosphere would contain 99% CO_2 if Earth had no life (that is, photosynthesizing microbes and plants), and were in thermodynamic equilibrium. CO_2 is a good absorber of longwave radiation. With the present energy output of the sun, a CO_2 atmosphere would retain the reradiated heat and maintain the Earth at a mean surface temperature of about 290° ± 50°C instead of the present 15°C. Venus and Mars, our neighboring planets, have an atmosphere containing 98 and 95% CO_2, respectively, and are thus unlikely to have life.

An extraterrestrial intelligence could detect life on Earth simply by pointing a spectrometer at the atmosphere of Earth because of the abundant presence of O_2 and N_2 and the small percentage of CO_2 are indications of nonequilibrium produced by life forms. There are many indications that the change in atmospheric CO_2 content has influenced climate. A recent authoritative description can be found in papers by Arthur[3] and Pollack,[4] which discuss climate in Earth history in the geological past. It appears from Pollack[4] that the main trend of the mean surface temperature of Earth can be described by a decrease over 3.5×10^9 years, but that within most of that period the temperature remained within ± 10°C of the present mean of 15°C. In order for this to happen, CO_2 concentrations had to decrease by two to three orders of magnitude in order to accommodate stable temperatures during a linear increase in solar luminosity of about 30% over 3.5×10^9 years. Photosynthesis and subsequent incorporation of organic material in rocks is supposed to have depleted atmospheric CO_2.[1,2] Arthur[3] mentions that in the boundary between the Cretaceous and the Tertiary period (about 66 million years before present), there was a sudden injection of 9×10^{19} g of CO_2 into the atmosphere-ocean system. This amount is two orders of magnitude larger than the accumulated production of CO_2 from burning of fossil fuel between the years 1860 and 1980 of 5×10^{17} g CO_2 reported by Watts.[5] On a geological time scale there appears to exist a powerful feedback mechanism which can stabilize CO_2 and temperature perturbations.

Air samples up to 40,000 years old can be found in air trapped in polar ice. Neftel et al.[6] have shown that CO_2 concentrations have not remained stable during that time span, and it appears that the atmospheric CO_2 concentration decreased from 300 μll^{-1} 40,000 years ago to between 150 and 200 μll^{-1} during the last ice age 20,000 years ago, and increased rapidly around 10,000 years ago to between 280 and 300 μll^{-1} as the Ice Age ended. According to Clark et al.[7] we do not know whether "the natural increase in CO_2 may have followed or accompanied, rather than preceded, periods of natural climate warming."

The agro-industrial activities of mankind have influenced the carbon budget of the world over only a relatively short time span. Land clearing for agriculture has taken place for 2000 to 3000 years in Southeast Asia, Mesopotamia, and the Mediterranean region on a limited scale, but only during the last few hundred years — with the population increase which followed the industrial revolution — has land clearing become a major factor in the global carbon budget. According to Clark et al.[7] the best estimates of current deforestation suggest a net carbon release of about 2×10^{15} g carbon year^{-1}. This is somewhat less than the amount released from burning of fossil fuel and cement manufacturing which, according to Wats,[5] accounts for 5.3×10^{15} g C year^{-1} or 1.9×10^{16} g CO_2 year^{-1}.

IV. THE INFLUENCE OF GLOBAL CO_2 ENRICHMENT ON CLIMATE

The burning of fossil fuel is presently the major source of energy in modern technology. Although some optional energy sources exist (atomic power, wind power and hydroelectric power), it is not realistic to assume that it will be possible to shift the energy technology on a massive scale in less than a century. It is therefore generally assumed that burning of fossil fuel will continue to be the most common energy source in the next century or even longer. Estimates of future energy consumption until the year 2000, expressed in percent yearly increases (Table 1 in Clark et al.[7]), is 2.94 ± 0.56 (mean of 15 estimates $\pm$ SD) while estimates for the next century usually are closer to 2% per year.

Not all the released CO_2 remains in the air. According to Clark et al.,[7] the most likely value for the airborne fraction is 0.4, while the remaining part is absorbed in the oceans.

According to Stanhill,[8,9] a group of measurements of CO_2 in the bulk air in France prior to 1890 shows that the CO_2 content was 281 ± 9 μll^{-1} in Montsouris and 284 ± 9 μll^{-1} at three other locations in France. From this concentration it rose to over 340 μll^{-1} in 1982 and the present rate of increase approaches 2 μll^{-1} per year, according to Keeling, Bacastow, and Whorf.[10] It is expected that a doubling of atmospheric CO_2 will take place around the middle or second half of the next century, and if all fossil fuel will be burned over the next few centuries, Clark et al.[7] estimate that the peak atmospheric CO_2 concentration would be slightly less than 2000 μll^{-1}. Most climate modelers estimate that each doubling of atmospheric CO_2 will increase the mean surface temperature of Earth by $2.3 \pm 1.6°C$ (mean and SD of 34 model forecasts) as described by Clark et al.[7] The 60 μll^{-1} CO_2 increase over the last 100 years has not caused any warming,[7] but if the models are correct we can expect a major warming to take place in the future. An important aspect of the forecasts is that most models expect the temperature increase at high latitudes to be a factor of two or three greater than the warming in the tropics.

A temperature rise in the high latitudes will increase oxidation of soils with high organic matter content. The amount of carbon in soil organic matter (1.2×10^{18}g) has been estimated, according to Watts,[5] to be about two thirds the total amount of carbon in the biosphere. It can therefore be expected that if the polar soils in the USSR, Canada, and elsewhere are warmed up, they will oxidize and add CO_2 to the atmosphere, which will act as a feedback to the atmospheric CO_2.

V. FEEDBACK FROM PLANTS

In contrast to the yet unobserved global warming, considerable knowledge about the effect of elevated CO_2 on plant production has been gathered. The main climatic effects of global CO_2 enrichment are, according to Schlesinger,[12] increases in temperature and precipitation. This is likely to increase net photosynthesis rate, change transpiration rate, and influence the water shortage of plants. References to about 800 papers on the direct effects of carbon dioxide on plants and plant communities can be found in Strain et al.[13]

Kimball[14] analyzed 770 experiments on CO_2 enrichment and found that a doubling of CO_2 increased yield in cotton by 104%, fruit crops by 11%, C_3 grains by 15%, legumes by 17%, root and tuber crops by 49%, woody plants by 40%, and dry matter of all C_4 species by 11%. Baker and Lambert[15] arrived at similar figures. Zelitch[16] showed that crop yield is closely related to the net photosynthetic assimilation of CO_2 throughout an entire season, and hence, to CO_2 concentration. A comprehensive description of plant reactions to elevated CO_2 in the global atmosphere can be found in a book edited by Lemon,[17] while CO_2 enrichment of greenhouse plants is the subject of proceedings edited by Mortensen[18] and of a book edited by Enoch and Kimball.[19]

The most abundant group of plants in existence today, the so-called C_3 plants, never fully adapted to the combination of high oxygen and low carbon dioxide concentration. Zelitch[16] refers to the 33 to 50% inhibition of photosynthesis of C_3 plants caused by the present 21% O_2 — compared with 2% O_2 — as the "oxygen stress". The inhibition of photosynthesis of C_3 plants is mitigated by elevated levels of CO_2. The C_4 plants, which are believed to have evolved later, show, according to Zelitch,[16] no sign of oxygen inhibition. The carbon assimilation of all plants (C_3, C_4, and CAM) subjected to an atmosphere with 21% O_2 is increased by additional CO_2 above the ambient 340 ull^{-1} level. We can therefore claim that all plants are subjected to "carbon dioxide deficiency stress".

The effect of a CO_2 increase on net photosynthesis rate of C_3 plants can be estimated, using an empirical model by Enoch and Sacks[20] based on data from an experiment by Enoch and Hurd[21] with 120 combinations of light, CO_2 concentrations, and temperatures. Using outdoor light conditions representative of the monthly mean of each hour at the 30, 40, 50, and 60° latitudes, the effects of CO_2 concentration on yearly net photosynthesis rate of C_3 plants were calculated. For the sake of simplicity, temperatures of 20°C during the day and 10°C at night were assumed. A doubling of ambient atmospheric CO_2 calculated by Enoch and Hurd[22] to enhance net photosynthetic rate of carnation plants by 35, 33, 28, and 25% on a yearly basis for the 60, 50, 40, and 30° latitudes, respectively. A similar estimation by Enoch[23] of the effect of the increase which took place in this century from 300 to 330 μll^{-1} CO_2, gave a 3% increase in yearly net photosynthesis rate for a location near Tel Aviv, Israel.

C_3 plants seem to be adapted to the increase in global temperature caused by increases in atmospheric CO_2 concentration. Most climatic models link a doubling of atmospheric CO_2 to a global temperature increase of 2.0 to 2.5°C, according to Schlesinger.[12] Carnations, a C_3 plant, increase the leaf temperature at which maximal net photosynthesis takes place by 2°C for each doubling in pCO_2 over the range 200 to 3100 μll^{-1}, according to Enoch and Hurd.[21]

In measurements of transpirational water loss from carnation plants, Enoch and Hurd[22] found that there was a linear decrease of 60% in transpiration between ambient and 1500 ull^{-1} CO_2. Extrapolation from this relationship, and assuming all other factors constant, we can calculate that the global increase in CO_2 has reduced plant transpiration during this century by 1.6%. According to Enoch and Hurd,[22] a further reduction of about 10% may be reached within the next 50 years. Such a change could in the future reduce water shortage of terrestrial plant communities as "transpiration in higher plants accounts for about three-quarters of the water that is vaporized at the global land surface and one-eighth of that vaporized over the entire globe", according to Cowan.[24]

In addition to calculating plant reactions to a global CO_2 enrichment, one can observe how the CO_2 concentration has been influenced by the combined primary production rate of all plant communities. Seasonal atmospheric CO_2 concentration — highest in May and lowest in October — have been interpreted as the influence of net photosynthesis of terrestrial plants by Keeling.[25] Other factors which could have influenced this seasonal amplitude, such as seasonal use of fossil fuel, have been estimated by Pearman and Hyson[26] to cause only minor variations.

VI. THE ROLE OF CO_2 CONCENTRATION IN THE DEVELOPMENT AND YIELD OF HIGHER PLANTS

The growth and development of higher plants are influenced by atmospheric CO_2 content. Elevated CO_2 concentrations tend to cause increased root-to-top ratio is higher plants. This was observed in different C_3 plants, like radish, by Sionit, Hellmers, and Strain;[27] wheat, by Sionit, Strain, and Hellmers;[28] and maize and other C_4 plants, by Rogers, Thomas, and Bingham.[29] An increase in CO_2 concentration also tends to increase stem weight as found, for instance, by Rogers et al.[30] for a soybean crop and by Funsch et al.[31] for *Pinus strobus* L. (white pine). Rooting of stem sections also appears to be increased by elevated levels of CO_2, as shown, for instance, by Molnar and Cumming[32] for *Chrysanthemum*.

In addition, it appears that branching and tillering are enhanced by elevated CO_2. This was shown by Imai and Murata[33] for barley and by Mortensen and Moe[34] for roses.

The growth rate of the whole plant also appears to be positively correlated with CO_2 concentration as observed, for instance, by Pfeufer and Krug[35] for spinach and by Enoch, Rylski, and Samish[36] for lettuce.

High CO_2 levels have also been reported by Calvert[37] and by Wittwer and Robb[38] to enhance the development of the inflorescence in tomato. The number of flowers formed on cucumber plants was doubled by a tenfold increase in CO_2 according to Enoch, Rylski, and Spigelman.[39] These are but a few examples which illustrate the general point that growth and development of plants are enhanced by increasing the atmospheric CO_2 concentration.

VII. CARBON DIOXIDE ENRICHMENT OF GREENHOUSES: ITS POTENTIAL AND CURRENT PRACTICES.

The first gas exchange experiment was made by Joseph Priestly[40] in 1772, (quoted from Rabinowitz[41]); he wrote "Finding that candles would burn very well in air in which plants had grown a long time . . . I thought it was possible that plants might also restore the air which had been injured by burning of candles. Accordingly, on the 17th of August 1771 I put a sprig of mint into a quantity of air in which a wax candle had burned out and found that on the 27th of the same month another candle burnt perfectly well in it." A quantitative approach to gas exchange of plants was described by Nicolas Theodore de Saussure[42] in 1804. He measured the increase in dry weight caused by the assimiliation of a certain quantity of CO_2 (quoted from Rabinowitz[41]).

A. Carbon Dioxide Uptake Efficiency (CDUE)

In this section, de Saussure's notion is developed further by examining, in some detail, what happens to a quantity of CO_2 released in a greenhouse with elevated CO_2 during a short time interval, e.g., a few minutes). The ratio of CO_2 assimiliated by the crop to the amount of CO_2 supplied to the greenhouse is called the carbon dioxide uptake efficiency (CDUE).

It is now possible to investigate the conditions for maintaining a CDUE value which, during each time interval is profitable, and to do this in four modes of CO_2 enrichment.

1. Constant CO_2 Supply Rate

A common method of CO_2 supply is to release compressed CO_2 at a constant rate or to use a CO_2 generator burning fossil fuel such as propane, butane, or kerosine at a fixed rate. In a greenhouse this gives a variable CO_2 concentration. When the light is low, crop net photosynthesis rate is low, and hence, CO_2 accumulates and high CO_2 levels are reached. Under these conditions, the CO_2 loss from the greenhouse to the atmosphere is large, i.e., in the early morning and later afternoon hours CDUE is low. During the middle of the day, at maximal solar radiation flux density, the maximal crop net photosynthesis rate is reached. If the constant CO_2 supply rate is smaller than the maximal crop net photosynthesis, CO_2 concentration decreases toward the middle of the day and may even drop lower than 340 ull^{-1}.

2. Constant CO_2 Concentration

If a constant but elevated CO_2 concentration is applied during the day, the rate of CO_2 loss to the atmosphere is roughly constant during the day (ignoring the influence of varying windspeed). However, the rate of CO_2 uptake still follows the amount of crop-intercepted solar radiation; hence, CDUE is smaller in the morning, increases toward midday, and decreases toward the end of the daylight hours. The production costs of one unit of additional yield thus change during the day and the integrated CDUE for the whole day is suboptimal.

3. Pulsed CO_2 Enrichment

During midday hours in many climatic regions with high solar radiation flux density, greenhouses reach a supraoptimal temperature level, if the greenhouse is not ventilated. However, when greenhouses are ventilated, the higher rate of CO_2 loss to the bulk atmosphere prevents the economic use of CO_2 enrichment. A partial solution to the problem of CO_2 enrichment during periods with high solar radiation flux densities is to use alternating periods of ventilation (typically 3 to 10 min) without CO_2 enrichment followed by periods in which the greenhouse is closed and enriched with CO_2 (typically 10 to 30 min), as described by Enoch.[43] During the ventilation period the temperature decreases from the upper permissible ventilation temperature (t_{vent}), to the outdoor air temperature (or close to it). Immediately after the greenhouse is closed it can be CO_2-enriched and the elevated CO_2 concentration can be maintained until the greenhouse is ventilated again. In order not to waste any CO_2, it is possible to enrich the closed greenhouse with an amount of CO_2 such that the crop can manage to assimilate it just before the greenhouse reaches the "ventilation temperature".

In order to calculate the amount of CO_2 the crop can assimilate before ventilation becomes necessary, it would be necessary to combine a quantitative dynamic simulation model of crop CO_2 uptake under nonsteady-state CO_2 concentrations with a dynamic heat balance model of greenhouses.

Bellamy and Kimball[44] have constructed a simulation model which calculates the percentage of time a greenhouse can become CO_2-enriched by the pulse method under different regions climatic conditions. It appears from their work that for most locations between the 60 and 32°N latitudes, an increase in yield due to CO_2 enrichment in pulses could raise the annual yield by up to 26%. The typical yield increase due to constant CO_2 enrichment at lower light levels and pulsed CO_2 enrichment at high solar radiation flux densities was calculated to be 50, 51, 48, 48, 47, and 36% for Oslo, De Bilt (Holland), Milano, Columbus (Ohio), Tokyo, and Tel-Aviv, respectively.

4. Constant CDUE

The simplest situation to describe is CO_2 enrichment of a crop in an airtight greenhouse. All supplementary CO_2 will eventually be assimilated by the crop and consequently CDUE = 1.

From the harvest index and the dry-to-fresh-weight ratio, a proportionality factor between assimilated CO_2 and fresh weight yield can be calculated. Since CDUE = 1, assimilated CO_2 equals CO_2 supplied to the greenhouse. If we know the price of the yield from enriched and nonenriched greenhouses, and the proportionality factor, we can calculate the break-even price for CO_2.

No commercial greenhouses are completely airtight. Most of them have an hourly air exchange rate of between 0.2 (for the best constructed) and 4 (for the leakiest structure).

The CO_2 lost through openings in the greenhouse is proportional to the CO_2 concentration difference between the inside and outside. Short-term gross-photosynthesis (and possibly yield) is, according to Enoch and Sacks,[20] for the first approximation proportional with the log of CO_2 concentration. The link between CO_2 concentration (and therefore costs) and diminishing returns makes it obvious that there must be an economically optimal CO_2 concentration.

We can determine how CO_2 concentration should change during the day in order to give us a constant CDUE value. For the steady state situation, total CO_2 supply to the greenhouse equals CO_2 assimilation + CO_2 loss by ventilation. All three factors can be measured independently, as outlined below.

B. Measurement of CDUE

The ratio between CO_2 assimilated during CO_2 enrichment (and retained in the crop) and the amount of CO_2 supplied to the greenhouse during CO_2 enrichment, can, according to Lake,[45] be measured over short time intervals by gas exchange measurement and over longer periods by ^{14}C and ^{13}C isotope analysis of the crop, as described by Enoch et al.[46]

Lake[45] measured CO_2 inside and outside the greenhouse, and the rate of air exchange from the greenhouse, with a tracer gas N_2O. The method provides information about CO_2 uptake over time intervals as short as 10 to 30 min. By measuring subsequent short periods, the CO_2 uptake pattern over days can be obtained.

If we could make detailed measurements of gas exchange from many greenhouses, including the whole growth season, greenhouse types, and CO_2 application methods, it would be possible to determine the method that would give the highest CDUE value. However, continuous gas exchange measurements are time consuming, expensive, and hence, not performed in more than a few greenhouse research centers in the world.

The grower who practices CO_2 enrichment generally knows how much CO_2 he has supplied to the greenhouse. The amount of CO_2 can be measured with a gas meter used for household gas in many countries. What he would like to know is, how much of the supplied CO_2 has been assimilated by the crop and retained until the day of harvest. This can be determined by carbon isotope analysis. Isotope analysis of CO_2 in bulk atmospheric air, and of CO_2 used for CO_2 enrichment, shows that the values differ in the relative abundance of ^{14}C, ^{13}C, and ^{12}C.

The abundance of ^{13}C in the global atmosphere CO_2 is 1.11% of the total amount of carbon, and that of ^{14}C is about 10^{-10}, according to Fritz and Fontes.[47] In CO_2 of fossil fuel origin, all the ^{14}C has decayed ($t_{1/2} = 5700$ years) and the ^{13}C concentration is approximately 2% less than in the atmosphere. The accurate measurement of these isotopes can be used to determine the relative contributions of CO_2 from the atmosphere and from fossil fuel origin to plant tissue. The procedure used by Enoch et al.[46] is a mass balance of the plants' isotopic content of carbon.

The isotope analysis was made on leaves of tomato plants grown in three greenhouses in Israel; one was used as control and received no additional CO_2, one was CO_2-enriched by releasing pure CO_2 from a container, and in one greenhouse, CO_2 was produced by burning propane and butane gas. CO_2 concentrations of 1100 ± 100 μll^{-1} were maintained during the daytime when the greenhouse could be closed without the temperature reaching 28°C. Under the climatic conditions in Israel, this occurred for about seven out of the ten daylight hours during midwinter. In measurements reported by Enoch et al.[46] it was shown by ^{14}C analysis that approximately 42% of the plant organic carbon comes from artificially added CO_2. Similar results were obtained from ^{13}C analysis when the pressure-dependent isotope separation factor was taken into account.

VIII. CONCLUSIONS

In this chapter various aspects of interactions between the atmospheric carbon dioxide content and plants have been described.

Over the eons, life forms have existed on Earth. Plants' carbon uptake and incorporation of the dead plants in sediments has caused a depletion of atmospheric CO_2 which has contributed to the maintenance of a suitable temperature regime for all biota.

In greenhouses, a constant elevated concentration and/or pulse-CO_2 enrichment is an effective method for increasing plant production by 30 to 50%. However, local conditions influencing the costs of CO_2 sources and the price of the yield may change the profit level. An outline for estimating the break-even point of CO_2 enrichment is given in the description of the carbon dioxide-uptake-efficiency concept.

Anthropogenic activities are changing the CO_2 content of the global atmosphere rapidly. The CO_2 increase from 280 to 340 μll^{-1} which has taken place during the last 100 years is likely to have been beneficial for plant productivity. Over the next century a further CO_2 increase to 600 μll^{-1} is expected. In the century after that, CO_2 may reach a maximum of 1000 to 2000 μll^{-1}. Future elevated CO_2 levels are likely to be beneficial for plant productivity and thereby increase mankind's supply of food, fiber, and fuel, unless the added atmospheric CO_2 causes a severe climate deterioration.

REFERENCES

1. **Margulis, L.,** *Symbiosis in Cell Evolution; Life and its Environment on Early Earth,* W. H. Freeman, San Francisco, 1981.
2. **Lovelock, J. E.,** *Gaia, A New Look at Life on Earth,* Oxford University Press, England, 1979.
3. **Arthur, M. A.,** The carbon cycle — controls on atmospheric CO_2 and climate in the geological past, in *Climate in Earth History,* Berger, W. H. and Crowell, J. C., Eds., National Academic Press, Washington, D.C., 1982, 55.
4. **Pollack, J. B.,** Astronomical and atmospheric effects on climate, in *Climate in Earth History,* Berger, W. H. and Crowell, J. C., Eds., National Academic Press, Washington, D.C , 1982, 68.
5. **Watts, J. A.,** The carbon dioxide question: data sampler, in *Carbon Dioxide Review: 1982,* Clark, W. C., Ed., Oxford University Press, New York, N.Y., 1982, 431.
6. **Neftel, A., Oeschger, H., Schwander, J., Stauffer, B., and Zumbrunn, R.,** Ice core sample measurements give atmospheric CO_2 content during the past 40000 years, *Nature,* 295, 220, 1982.
7. **Clark, W. C., Cook, K. H., Marland, G., Weinberg, A. M., Rotty, R. M., Bell, P. R., Allison, L. J., and Cooper, C. L.,** The carbon dioxide question: perspectives for 1982, in Carbon Dioxide Review: 1982, Clark W. C., Ed., Oxford University Press, New York, N.Y., 1982.
8. **Stanhill, G.,** The Montsouris series of carbon dioxide concentration measurements, 1817—1910, *Climatic Change,* 4, 221, 1982.
9. **Stanhill, G.,** A further reply concerning the accuracy of the Montsouris series of carbon dioxide concentration measurements, 1877—1910, *Climatic Change,* 6, 413, 1984.
10. **Keeling, C. D., Bacastow, R. B. N., and Whorf, T. P.,** Measurements of the concentration of carbon dioxide at Mauna Loa Observatory, Hawaii, in *Carbon Dioxide Review: 1982,* Clark, W. C., Ed., Oxford University Press, New York, N.Y., 1982, 377.
11. **Waggoner, P. E.,** Agriculture and a climate changed by more carbon dioxide, in *Changing Climate, Report of the Carbon Dioxide Assessment Committee,* National Academy Press, Washington, D.C., 1983, 383.
12. **Schlesinger, M. E.,** Simulating CO_2-induced climate change with mathematical climate models: capabilities, limitations and prospects, in Proceedings: Carbon Dioxide Research Conference: Carbon Dioxide, Science and Consensus III, Pub. No. 021, U.S. Department of Energy, Carbon Dioxide Research Division, Washington, D.C., 1983, 3.
13. **Strain, B. R., Sionit, N., and Cure, J.,** Direct Effects of Carbon Dioxide on Plants and Plant Communities: A Bibliography, Department of Botany, Duke University, Durham, N.C., 1984, 70.

14. **Kimball, B. A.,** Influence of elevated CO_2 on crop yield, in *Carbon Dioxide Enrichment of Greenhouse Crops,* Vol. II, Enoch, H. Z. and Kimball, A. B., CRC Press, Boca Raton, Fla., 1986, chap. 8.

15. **Baker, D. N. and Lambert, J. R.,** The analysis of crop responses to enhanced atmospheric CO_2 levels, in Workshop on Environmental and Societal Consequences of a Possible CO_2-Induced Climate Change, Publ. No. 009, U.S. Department of Energy, Carbon Dioxide Research Division, Washington, D.C., 1980, 275.

16. **Zelitch, I.,** The close relationship between net photosynthesis and crop yield, *BioScience,* 32, 796, 1982.

17. **Lemon, E. R., Ed.,** *CO_2 and Plants: The Response of Plants to Rising Levels of Atmospheric Carbon Dioxide,* Westview Press, Boulder, Colo., 1983, 280.

18. **Mortensen, L., Ed.,** Symposium on CO_2 enrichment, *Acta Hortic.,* 162, 1984.

19. **Enoch, H. Z. and Kimball, B. A., Eds.,** *Carbon Dioxide Enrichment of Greenhouse Crops,* CRC Press, Boca Raton, Fla., 1986.

20. **Enoch, H. Z. and Sacks, J. M.,** An empirical model of CO_2 exchange of a C_3 plant in relation to light, CO_2 concentration and temperature, *Photosynthetica,* 12, 150, 1978.

21. **Enoch, H. Z. and Hurd, R. G.,** Effect of light intensity, carbon dioxide concentration and leaf temperature on gas exchange of spray carnation plants, *J. Exp. Bot.,* 84, 1977.

22. **Enoch, H. Z. and Hurd, R. G.,** The effect of elevated CO_2 concentrations in the atmosphere on plant transpiration and water use efficiency: a study with potted carnation plants, *Int. J. Biometerol.,* 23, 343, 1979.

23. **Enoch, H. Z.,** Diurnal and seasonal variations in the carbon dioxide concentration of the lower atmosphere in the coastal plain of Israel, *Agric. Meterol.,* 18, 373, 1977.

24. **Cowan, I. R.,** Water use in higher plants, in *Water, Planets, Plant and People,* McIntyre, A. K., Ed., Australian Academy of Science, Canberra, 1978, 71.

25. **Keeling, C. D.,** the Global Carbon Cycle: What we Know and Could Know from Atmospheric, Biospheric and Oceanic Observations, in Proc. Carbon Dioxide Research Conference: Carbon Dioxide, Science and Consensus. II. Publication No. 021., U.S. Department of Energy, Carbon Dioxide Research Division, Washington, D.C., 1983, 3.

26. **Pearman, G. I. and Hyson, P.,** The annual variation of atmospheric CO_2 concentration observed in the Northern Hemisphere, *J. Geo-Phys. Res.,* 86, 9839, 1981.

27. **Sinoit, N., Hellmers, H., and Strain, B. R.,** Interactions of atmospheric CO_2 enrichment and irradiance on plant growth, *Agron. J.,* 74, 721, 1982.

28. **Sionit, N., Strain, B. R., and Hellmers, H.,** Effects of different concentrations of atmospheric CO_2 on growth and yield components of wheat, *J. Agric. Sci.,* 79, 335, 1981.

29. **Rogers, H. H., Thomas, J. F., and Bingham, G. E.,** Response of agronomic and forest species to elevated atmospheric carbon dioxide, *Science,* 220, 42B, 1983.

30. **Rogers, H. H., Bingham, G. E., Cure, J. E., Heck, W. W., Heagle, A. S., Israel, D. W., Smith, J. M., Surano, K. A., and Thomas, J. F.,** Response of Vegetation to Carbon Dioxide: Field Studies of Plant Responses to Elevated Carbon Dioxide Levels, Report No. 001, U.S. Department of Energy, Carbon Dioxide Division, Washington, D.C., 1980.

31. **Funsch, R. W., Mattsoa, R. H., and Mowry, G. R.,** CO_2-supplemented atmosphere increases growth of *Pinus strobus* seedlings, *For. Sci.,* 16, 459, 1970.

32. **Molnar, J. M. and Cumming, W. A.,** Effect of carbon dioxide on propagation of softwood conifer and herbaceous cuttings, *Can. J. Plant Sci.,* 48, 595, 1968.

33. **Imai, K. and Murata, Y.,** Effect of carbon dioxide concentration on growth and dry matter production of plants. I. Effects of leaf area, dry matter, tillering dry matter distribution ratio and transpiration, *JPN. J. Crop Sci.,* 45, 59B, 1976.

34. **Mortensen, L. M. and Moe, R.,** Growth responses of some greenhouse plants to environment. V. Effects of CO_2, O_2 and light on net photosynthetic rate in *Chrysanthemum morifolium* R. *Sci. Hortic.,* 19, 141, 1983.

35. **Pfeufler, B. and Krug, H.,** Effects of high CO_2-concentrations on vegetables, *Acta Hortic.,* 162, 37, 1984.

36. **Enoch, H. Z., Rylski, I., and Samish, Y.,** CO_2 enrichment to cucumber, lettuce and sweet pepper plants grown in low plastic tunnels in a subtropical climate, *Isr. J. Agric. Res.,* 20, 63, 1970.

37. **Calvert, A.,** Effects of day and night temperatures and carbon dioxide enrichment on yield of glasshouse tomatoes, *J. Hortic. Sci.,* 47, 231, 1972.

38. **Wittwer, S. H. and Robb, W.,** Carbon dioxide enrichment of greenhouse atmospheres for crop production, *Econ. Bot.,* 18, 34, 1964.

39. **Enoch, H. Z., Rylski, I., and Spigelman, M.,** CO_2 enrichment of strawberry and cucumber plants grown in unheated greenhouses in Israel, *Sci. Hortic.,* 5, 33, 1976.

40. **Priestly, J.,** *Philos. Trans. R. Soc. London,* 147, 62, 1772, (quoted from Rabinowitz.[41])

41. **Rabinowitz, E. I.,** *Photosynthesis and Related Processes,* Interscience, New York, 1945, 13.
42. **de Saussure, N. T.,** Recherches chimiques sur la vegetation, Nyon, Paris, 1804 (quoted from Rabinowitz, E. I.) in *Photosynthesis and Related Processes,* Interscience, New York, 1945, 23.
43. **Enoch, H. Z.,** Carbon dioxide uptake efficiency in relation to crop intercepted solar radiation, *Acta Hortic.,* 162, 137, 1984.
44. **Bellamy, L. A. and Kimball, B. A.,** Carbon dioxide encirhment duration and heating credit as determined by climate, in *Greenhouse Crops,* Enoch. H. Z. and Kimball, B. A., Eds., CRC Press, Boca Raton, Fla., 1986.
45. **Lake, J. V.,** Measurement and control of carbon dioxide assimilation by glasshouse crops, *Nature (London),* 209, 97, 1966.
46. **Enoch, H. Z., Carmi, I., Rounick, J. S. and Magaritz, M.,** Use of carbon isotopes to estimate incorporation of added CO_2 by greenhouse-grown tomato plants, *Plant Physiol.,* 76, 1083, 1984.
47. **Fritz, P. and Fontes, J. Ch.,** Introduction, in *Handbook of Environmental Isotope Geochemistry,* Fritz, P. and Fontes, J. Ch., Eds., Elsevier, Amsterdam. 1979, 4.

INDEX

Printed and bound by CPI Group (UK) Ltd, Croydon, CR0 4YY

22/10/2024

01777630-0019